WITHDRAWN

# Plant Proteolytic Enzymes

## Volume II

Editor

**Michael J. Dalling, Ph.D.**
Senior Lecturer
School of Agriculture and Forestry
The University of Melbourne
Parkville, Victoria, Australia

CRC Press, Inc.
Boca Raton, Florida

**Library of Congress Cataloging-in-Publication Data**
Main entry under title:

   plant proteolytic enzymes
   Bibliography: p.
   Includes index.
   1. Proteolytic enzymes.   2. Botanical chemistry.
I. Dalling, Michael J.
QK898.P82P55   1986     581.19′256     85-24319
ISBN 0-8493-5682-2 (v. 1)
ISBN 0-8493-5683-0 (v. 2)

   This book represents information obtained from authentic and highly regarded sources. Reprinted material is quoted with permission, and sources are indicated. A wide variety of references are listed. Every reasonable effort has been made to give reliable data and information, but the author and the publisher cannot assume responsibility for the validity of all materials or for the consequences of their use.

   All rights reserved. This book, or any parts thereof, may not be reproduced in any form without written consent from the publisher.

   Direct all inquiries to CRC Press, Inc., 2000 Corporate Blvd., N.W., Boca Raton, Florida, 33431.

© 1986 by CRC Press, Inc.
International Standard Book Number 0-8493-5682-2 (v.1)
International Standard Book Number 0-8493-5683-0 (v.2)

Library of Congress Card Number 85-24319
Printed in the United States

# PREFACE

Knowledge on protein degradation and on the proteolytic enzymes involved appears to be equally important for the understanding of cell metabolism and development and knowledge on protein synthesis. This is not an overstatement. It is simply concluded from the fact that protein degradation and protein synthesis depend on each other. Both processes contribute equally to the vitally important protein metabolism, both of them are equally involved in protein turnover and in all of the changes in quantity and quality of cellular protein as associated with cell differentiations and, hence, with development. Leaf senescence provides just one example demonstrating that even the marked loss of protein depends on protein synthesis. Seed germination may be mentioned as an example for the opposite case in which protein synthesis is intimately associated and even dependent on proteolysis.

Biologists have been much more interested and successful in the elucidation of protein synthesis as compared with protein degradation which has been, and to some extent still is, a rather neglected field. This is certainly due to the fact, that protein synthesis was appreciated as the key to the understanding of the connection between genome and metabolism. However, the elucidation of protein synthesis may have been easier than the work toward the understanding of protein degradation although the latter represents, in biochemical terms, nothing but simple hydrolysis.

A considerable number of plant proteolytic enzymes has been described so far, but only in rare cases have the catalysts been associated with a clear cut function. It is a truism that the number of proteases (including the number of proteases which will be discovered in the future) is much smaller than the number of protein species, i.e., the potential substrates present in plant cells. The million dollar question concerns, therefore, the explanation of specificity of protein degradation which has been observed in various instances, e.g., when turnover rates were determined for individual protein species or losses of certain proteins were followed in senescent leaves. It appears to be impossible to explain such phenomena with activities of highly specific proteolytic enzymes responsible for selective degradation of certain proteins. It rather seems that a comparatively low number of proteolytic enzymes is responsible for unspecific degradations of proteins with which they get into contact. Hence, the mechanisms responsible for the contact between proteolytic enzymes and those proteins which are destined for degradation appear to be an important aspect of protein degradation.

A hypothetical solution of this problem is subcellular compartmentation. The concept of lysosomes, originally developed for rat liver cells, was adopted for plant cells. In the past few years it has been documented convincingly that plant cells, indeed, contain a lysosomal extraplasmatic compartment, the vacuole, in which proteolytic enzymes together with other hydrolases are located. The concept of lysosomes is intelligible because proteolytic enzymes appear to be separated in a specific compartment, the cytoplasm, the truly living entity of the cell, being protected against the uncontrolled attack by digestive enzymes. Yet, the concept certainly does not explain the mechanism of selective and controlled protein degradation. It merely switches over from the specificity of proteolytic enzymes to the specificity of transport of cytoplasmic proteins into the vacuoles. In fact, the role of vacuolar porteolytic enzymes has so far been demonstrated unambiguously only in the very special case of protein bodies, in which the compartmentalization of the substrates, the reserve proteins, coincides with that of the proteolytic enzymes responsible for intracellular protein digestion during seed germination.

It is quite possible and even probable that only a fraction of proteolytic enzymes which can be assessed by using the conventional substrates has been discovered so far. It is not unlikely that proteolytic enzymes which are more directly related to the degradation of cytoplasmic proteins than are the vacuolar enzymes remained undiscovered because they have comparatively low activities and unusual properties or specificities that do not allow the determination with the common substrates. Recent findings of proteolytic activities in chloroplasts and in mitochondria support this view. The example of the proteolytic system of yeast (not covered in the present volume) may show the prerequisites for discovering minor proteolytic enzymes. At the same time the example demonstrates the limitations of work with higher plants.

Bakers yeast cells contain two principal endopeptidases (A and B), two carboxypeptidases (Y—formerly protease C and S), and several aminopeptidases. This proteolytic system is located in the vacuoles (A, B, Y, S, and two aminopeptidases). It is particularly interesting that yeast cells also contain several proteins which specifically inhibit the proteases A, B, and C. These inhibitor proteins are located in the cytosol. Hence, the proteolytic machinery in the vacuoles is fully active with the cytoplasm being protected, not only through compartmentation, but also by virtue of protease inhibitor proteins. Assuming that this system is responsible for protein degradation, one would expect that proteinase deficient mutants are not viable. Yet, mutants lacking the two vacuolar endopeptidases are able to grow, differentiate, and even sporulate, although the rate of protein turnover is markedly lower than in normal strains. Working with normal strains, Wiemken was able to demonstrate unambiguously that vacuoles are "the sole compartments" of endopeptidases.[1]

Working with mutants deficient in the vacuolar proteases, Wolf has not only discovered several novel proteolytic enzymes the activity of which in normal strains is completely masked; he also showed that "vacuoles are not the sole compartment of proteolytic enzymes in yeast";[2] the newly discovered proteinases D and E were found to be located outside the vacuoles. These extravacuolar enzymes may play essential roles in proteolysis and it is even feasible that the functions will be elucidated should selection for corresponding mutants be practicable. In any case, the researcher dealing with yeast is in a much more promising situation that his colleague working with higher plants who will have to wait for the availability of corresponding genitial research tools.

The present volume is undoubtedly a most valuable source of information about plant proteolytic enzymes. It covers not only the enzymological aspects, but also the various functions including those which are hypthetical at the moment and probably also difficult to prove in the future. Hopefully, the book will stimulate plant physiologists to step into the fascinating field of protein degradation and help to overcome the difficulties in understanding how the proteases are integrated in the metabolism of the living plant cells.

Michael J. Dalling

## REFERENCES

1. **Wienken, A., Schellberg, M., and Urech, K.,** Vacuoles: The sole compartments of digestive enzymes in yeast *(Saccharonyces cerevisiae)* ?, *Arch. Microbiol.*, 123, 23, 1979.
2. **Emter, O. and Wolf, D. H.,** Vacuoles are not the sole compartments of proteolytic enzymes in yeast, *FEBS Lett.*, 166, 321, 1984.

## THE EDITOR

**Michael J. Dalling, Ph.D.,** is a Reader in Crop Physiology in the School of Agriculture and Forestry at the University of Melbourne. A native of Australia, he received his B.Agr.Sc. from the University of Melbourne in 1967 and his M.Agr.Sc. from the same university in 1969. Dr. Dalling received his Ph.D. in Agronomy in 1971 from the University of Illinois. In 1974, he was appointed a Lecturer in Agronomy at the University of Melbourne and in 1985 he was appointed to the position of Reader in Crop Physiology. In 1981 he was a Senior Scholar of the Australian-American Education Foundation at the University of California at Davis and in 1981/82 he was a Visiting Professor in the Department of General Botany at the Swiss Federal Institute of Technology. Dr. Dalling is the author of numerous journal articles and is a member of the Australian Society of Plant Physiologists, Agronomy Society of Australia, The Wheat Breeding Society of Australia, and the American Society of Plant Physiologists. His current research includes an attempt to understand the process of plant senescence and how it determines or influences grain yield.

# CONTRIBUTORS

**Alan J. Barrett, Ph.D., Sc.D.**
Department of Biochemistry
Strangeways Laboratory
Worts Causeway
Cambridge, England

**Prem Lata Bhalla, Ph.D.**
Research Fellow
School of Agriculture and Forestry
The University of Melbourne
Parkville, Victoria, Australia

**Thomas Boller, Ph.D.**
Lecturer
Botanisches Institut
Abteilung Pflanzenphysiologic
University of Basel
Basel, Switzerland

**William R. Ellis, Ph.D.**
Director of Inoculation Research
Research Seed, Inc.
St. Joseph, Missouri

**Urs Feller, Ph.D.**
Privatdozent
Institute of Plant Physiology
University of Bern
Bern, Switzerland

**Arthur W. Galston, Ph.D.**
Professor
Department of Biology
Yale University
New Haven, Connecticut

**Ravindar Kaur-Sawhney, Ph.D.**
Research Associate
Department of Biology
Yale University
New Haven, Connecticut

**Heinrich Kauss, Ph.D.**
Professor
Department of Biology
University of Kaiserslautern
Kaiserslautern, West Germany

**R. Bruce Knox, Ph.,D., D.Sc.**
Professor
Department of Botany
Plant and Cell Biology Research Center
The University of Melbourne
Parkville, Victoria, Australia

**James E. Kruger, Ph.D.**
Research Scientist
Grain Research Laboratory
Canadian Grain Commission
Winnipeg, Manitoba, Canada

**John Michael Lord, Ph.D.**
Reader
Department of Biological Sciences
University of Warwick
Coventry, England

**Juhani Mikola, Ph.D.**
Associate Professor
Department of Biology
University of Jyväskylä
Jyväskylä, Finland

**Leena Mikola, Ph.D.**
Research Associate
Department of Biology
University of Jyväskylä
Jyväskylä, Finland

**Angela M. Nettleton**
School of Agriculture and Forestry
The University of Melbourne
Parkville, Victoria, Australia

**Ann Oaks, PhD.**
Professor
Department of Biology
McMaster University
Hamilton, Ontario, Canada

**K. R. Preston, Ph.D.**
Grain Research Laboratory
Canadian Grain Commission
Winnipeg, Manitoba, Canada

**Paul Reibach, Ph.D.**
Plant Physiologist
Rohm and Haas Research Laboratories
Spring House, Pennsylvania

**Colin Robinson, Ph.D.**
Lecturer
Department of Biological Sciences
University of Warwick
Coventry, England

**Mohan B. Singh, Ph.D.**
Research Fellow
Department of Botany
Plant Cell Biology Research Center
The University of Melbourne
Parkville, Victoria, Australia

**Richard Storey, Ph.D.**
Associate Professor of Biology
Department of Biology
The Colorado College
Colorado Springs, Colorado

**Carroll P. Vance, Ph.D.**
Professor and Research Plant Physiologist
U.S. Department of Agriculture
Agricultural Research Service
Department of Agronomy and Plant
 Genetics
University of Minnesota
St. Paul, Minnesota

**Fred W. Wagner, Ph.D.**
Professor
Department of Agricultural Biochemistry
University of Nebraska-Lincoln
Lincoln, Nebraska

**William Wallace, Ph.D.**
Senior Lecturer
Department of Agricultural Biochemistry
Waite Agricultural Research Institute
University of Adelaide
Glen Osmond, South Australia

**Karl A. Wilson, Ph.D.**
Associate Professor
Department of Biological Sciences
State University of New York at
 Binghamton
Binghamton, New York

TABLE OF CONTENTS

Chapter 1
Mobilization of Monocot Protein Reserves During Germination.......................... 1
**K.R. Preston and J.E. Kruger**

Chapter 2
Role of Proteolytic Enzymes in the Mobilization of Protien Reserves in the Germinating
Dicot Seed ................................................................................. 19
**Karl A. Wilson**

Chapter 3
Proteolytic Enzymes in Relation to Leaf Senescence ................................... 49
**Urs Feller**

Chapter 4
Role of Proteolytic Enzymes in the Post-Translational Modification of Proteins........ 69
**J. Michael Lord and Colin Robinson**

Chapter 5
Role of Proteinases in the Regulation of Nitrate Reductase............................ 81
**W. Wallace and A. Oaks**

Chapter 6
A Membrane-Derived Proteinase Capable of Activating a Galactosyl-Transferase Involved
in Volume Regulation of *Poterioochromonas* ........................................... 91
**Heinrich Kauss**

Chapter 7
Proteolytic Enzymes of Legume Nodules and Their Possible Role During Nodule
Senescence ................................................................................ 103
**C.P. Vance, P.H. Reibach, and W.R. Ellis**

Chapter 8
Chloroplast Senescence and Proteolytic Enzymes ..................................... 125
**Michael J. Dalling and Angela M. Nettleton**

Index ..................................................................................... 155

Chapter 1

# MOBILIZATION OF MONOCOT PROTEIN RESERVES DURING GERMINATION

### K. R. Preston and J. E. Kruger

## TABLE OF CONTENTS

| | | |
|---|---|---|
| I. | Introduction | 2 |
| II. | Nature of Storage Proteins | 2 |
| | A. Formation of Storage Protein Bodies During Maturation | 2 |
| | B. Properties of Storage Proteins | 3 |
| |     1. Albumins and Globulins | 3 |
| |     2. Prolamins | 3 |
| |     3. Glutelins | 5 |
| III. | Storage Protein Hydrolysis During Germination | 6 |
| | A. Barley, Wheat, and Rye | 7 |
| | B. Oats | 8 |
| | C. Rice | 9 |
| | D. Maize and Sorghum | 9 |
| IV. | Role of Proteolytic Enzymes in Storage Protein Hydrolysis | 10 |
| | A. Barley, Wheat, and Rye | 10 |
| | B. Maize and Sorghum | 12 |
| | C. Rice | 13 |
| References | | 13 |

## I. INTRODUCTION

In germinating seeds, the mobilization of storage proteins during germination provides a primary source of amino acids: nitrogen and carbon skeletons to the growing plant. In order to obtain a better understanding of this process, three basic ingredients are necessary. First, an understanding of the properties of storage proteins including their mode of deposition and their physical and chemical properties is required. Second, an understanding of the pattern of breakdown of these proteins during germination is necessary. And last, an understanding of the proteolytic enzyme systems responsible in terms of their mode of action is required. In the present chapter, the details of this process in terms of the above basic ingredients will be discussed for monocot seeds. Of necessity, due to the information available, this discussion has been restricted to monocot cereals.

## II. NATURE OF STORAGE PROTEINS

### A. Formation of Storage Protein Bodies During Maturation

The major sites of storage protein deposition in resting seeds occur in the form of protein bodies. In monocot seeds the bulk of the storage protein is deposited in the form of protein bodies in the endosperm, and to a lesser extent in the aleurone, during seed maturation. At present there is some confusion as to the mechanism involved in the formation of these protein bodies. In the aleurone of cereals such as maize, protein bodies appear to have a vacuolar origin.[1] However, in cereal endosperm tissue, evidence for both the formation of protein bodies in vacuoles and from the enlargement of regions of the rough endoplasmic reticulum (RER) has been presented. In maize, strong evidence suggests that storage proteins are synthesized on the RER and deposited in the lumen in the form of protein bodies surrounded by single membranes.[1-5] Formation of protein bodies in rice also appears to occur by a similar process.[6] However, in wheat and barley the situation isn't as clear.

Although it is now generally agreed that storage protein synthesis of wheat and barley storage proteins occurs on RER, the manner in which the storage proteins are deposited has been a subject of controversy. Studies by Miflin et al.[5] based upon the distribution of "marker" enzymes in isolated wheat and barley storage protein bodies suggests deposition of these bodies in the lumen of RER following synthesis. However, studies by Cameron-Mills and von Wettstein[7] suggest a vacuolar origin for protein bodies in barley. Campbell et al.[8] studied the development of intact plants and detached ears of wheat in culture by light and electron microscopy. They found accumulation of protein inside the cistern that led to the formation of protein bodies in the lumen of RER similar to those of maize.[3] However, protein bodies, though in smaller amounts, were also found in vacuolar structures. This led to a postulate that protein synthesis was carried out by polyribosomes attached to the RER and that proteins either accumulated in the lumen or were transported to vacuoles. The process appeared to occur preferentially during the early stages of development, while the latter process occurs preferentially during the later stage of development. Studies by Buttrose[9] are consistent with this interpretation. Recent studies by Bechtel et al.[10] suggest that the major site of protein accumulation in developing wheat is in the vacuoles via a soluble mode of transport.

Although the exact mode of protein body formation in cereal endosperms is subject to controversy, it is generally agreed that the protein bodies, following formation, are surrounded by a membrane. In mature seeds of maize[11] and millet,[12] discrete protein bodies are evident, while in wheat[12] and barley[11] they are not. On the basis of this evidence and of the resistance of isolated mature maize protein bodies to hydrolysis by proteinase-k in contrast to the susceptibility of wheat and barley protein bodies, Miflin and Burgess[13] suggested that the membrane surrounding maize protein bodies is intact after maturation, while those of barley and wheat are not. Thus the ability of proteolytic enzymes during germination to hydrolyze the storage proteins in various cereals may be, to some extent, dependent upon the state of the membrane surrounding the protein bodies.

### B. Properties of Storage Proteins

Cereal proteins are usually classified by their extractability in various solutions according to procedures similar to those originally developed by Osborne.[14-16] Sequential extraction of ground seeds or endosperms with saline solution, aqueous alcohol solutions, and dilute acid or basic solutions yield four distinct protein groups including albumins (water and salt soluble), globulins (salt-soluble; water-insoluble), prolamins (aqueous alcohol-soluble), and glutelins (acid or alkali-soluble). The relative proportions of each of these protein fractions can vary widely from species to species and from cultivar to cultivar within any species. In addition, variations in extraction procedure can have a strong influence.[17]

*1. Albumins and Globulins*

The salt-soluble albumins and globulins of cereals are generally considered to be cytoplasmic proteins. They consist largely of enzymes and of enzyme inhibitors.[19] Inhibitors of animal, microbial, and insect alpha-amylase and protease, which probably act as protective agents, can account for a large proportion of these proteins.[19-21] Inhibitors of endogenous enzymes have also been shown to be present in these protein fractions. This topic will be discussed in more detail later in the text. Although cereal albumins and globulins are not normally considered as storage proteins, they may serve this purpose as a secondary function following germination. The rapid disappearance of protease inhibitor activity in barley during germination is suggestive of this.[22]

Although the salt-soluble albumins and globulins of cereals account for the majority of the protein nitrogen in the germ and aleurone, they generally make up a much lower proportion of the endosperm protein. In wheat, salt-soluble proteins account for approximately 15 to 25% of endosperm nitrogen of which 70 to 75% are albumins.[23-27] Rye probably has the highest proportion of albumins plus globulins of any cereal, ranging from 40 to 50%.[23,24,28] Barley salt-soluble proteins appear to account for approximately 15 to 30% of total endosperm protein,[23,24,29,30] although lower values have been reported.[14,31] Values for rice[32,33] and oats[23,24,34] appear similar to barley. In maize, albumins and globulins have been reported to account for 7 to 20% of total seed or endosperm protein,[23,24,35] while in sorghum reported values have ranged from 20 to 26%.[23,36] In most of the studies cited above, albumins accounted for a much higher proportion of the salt-soluble proteins than did globulins.

For all species, both the albumin and globulin fractions are made up of large numbers of individual components as determined by electrophoresis.[25,30,37,38] Molecular weights vary widely, although the albumins generally give lower values than the globulins.[30,39,40]

Amino acid composition of cereal albumin and globulins are given in Table 1 and Table 2. Both groups of proteins have similar patterns including a high level of glutamic acid (plus glutamine), aspartic acid (plus asparagine), glycine, and alanine. The globulins generally have higher levels of arginine and lower levels of aspartic acid (plus aspargine) compared to the albumins.

*2. Prolamins*

Prolamins are normally considered to be the major group of proteins in cereal endosperm with the exception of oats and rice where glutelins are predominant. Pernollet[41] has summarized data concerning the distribution of protein fractions in protein bodies of cereal endosperms. In barley, wheat, maize, and sorghum, prolamins appear to account for 80 to 100% of the protein in these bodies, while in rice glutelins appear to predominate.

Cereal prolamins can sometimes be extracted with water,[42] but have a strong tendency to

## Table 1
## AMINO ACID COMPOSITIONS OF CEREAL ALBUMINS (mol %)[a]

|  | Wheat | Rye | Barley | Maize | Sorghum | Oats | Rice |
|---|---|---|---|---|---|---|---|
| Aspartic acid | 9.9 | 9.0 | 10.5 | 17.0 | 11.2 | 10.5 | 10.1 |
| Threonine | 3.7 | 3.9 | 4.6 | 4.3 | 4.9 | 4.3 | 4.5 |
| Serine | 5.8 | 5.8 | 6.1 | 5.8 | 5.9 | 8.4 | 6.1 |
| Glutamic acid | 21.4 | 22.6 | 14.3 | 12.6 | 12.4 | 12.7 | 14.5 |
| Proline | 9.5 | 12.3 | 7.7 | 8.7 | 5.2 | 6.3 | 4.7 |
| Glycine | 7.1 | 6.7 | 10.0 | 9.9 | 10.2 | 12.9 | 10.0 |
| Alanine | 7.1 | 6.6 | 8.5 | 10.2 | 10.7 | 7.8 | 9.6 |
| 1/2 Cystine | 3.3 | 2.4 | 3.9 | 1.8 | 1.4 | 7.0 | 1.9 |
| Valine | 5.7 | 4.9 | 6.0 | 4.5 | 6.1 | 4.5 | 6.0 |
| Methionine | 1.6 | 1.3 | 2.1 | 1.1 | 2.0 | 1.2 | 1.7 |
| Isoleucine | 3.2 | 3.3 | 3.2 | 2.9 | 3.2 | 2.6 | 3.5 |
| Leucine | 6.6 | 6.4 | 6.1 | 5.2 | 6.7 | 5.7 | 7.1 |
| Tyrosine | 2.9 | 2.4 | 3.5 | 3.9 | 3.1 | 3.3 | 3.2 |
| Phenylalanine | 3.2 | 4.0 | 2.7 | 2.0 | 3.2 | 2.8 | 3.3 |
| Lysine | 3.1 | 3.0 | 4.4 | 4.0 | 5.7 | 4.6 | 5.1 |
| Histidine | 1.8 | 1.7 | 1.8 | 2.1 | 2.3 | 1.6 | 2.4 |
| Arginine | 4.1 | 3.7 | 4.6 | 4.0 | 5.8 | 3.8 | 6.3 |

[a] Data of Wieser, H., Seilmeier, W., and Belitz, H., *Z. Lebensm. Unters. Forsch*, 170, 17, 1980. With permission.

## Table 2
## AMINO ACID COMPOSITIONS OF CEREAL GLOBULINS (mol %)[a]

|  | Wheat | Rye | Barley | Maize | Sorghum | Oats | Rice |
|---|---|---|---|---|---|---|---|
| Aspartic acid | 7.9 | 7.0 | 8.8 | 9.3 | 8.0 | 8.1 | 6.7 |
| Threonine | 4.5 | 4.5 | 4.7 | 5.1 | 4.4 | 4.2 | 2.8 |
| Serine | 6.2 | 6.5 | 6.1 | 7.0 | 7.6 | 6.5 | 6.6 |
| Glutamic acid | 15.6 | 17.4 | 13.2 | 11.0 | 12.4 | 16.4 | 14.9 |
| Proline | 7.1 | 8.0 | 7.0 | 5.7 | 5.3 | 5.4 | 5.7 |
| Glycine | 8.5 | 8.9 | 9.7 | 10.6 | 9.5 | 9.6 | 10.4 |
| Alanine | 7.7 | 7.8 | 8.5 | 11.0 | 9.9 | 7.6 | 8.2 |
| 1/2 Cystine | 3.7 | 2.1 | 3.1 | 3.3 | 3.6 | 2.4 | 4.2 |
| Valine | 6.4 | 6.0 | 6.4 | 5.9 | 6.0 | 6.1 | 5.8 |
| Methionine | 2.0 | 1.5 | 1.4 | 1.5 | 0.9 | 1.3 | 4.5 |
| Isoleucine | 3.6 | 3.7 | 3.0 | 3.9 | 3.4 | 3.9 | 2.5 |
| Leucine | 7.5 | 7.1 | 7.7 | 6.6 | 7.0 | 7.2 | 6.3 |
| Tyrosine | 3.0 | 2.3 | 2.8 | 2.7 | 3.1 | 2.8 | 3.8 |
| Phenylalanine | 3.2 | 3.7 | 3.4 | 3.3 | 3.4 | 4.1 | 2.9 |
| Lysine | 4.1 | 4.4 | 4.8 | 4.7 | 4.1 | 4.5 | 2.8 |
| Histidine | 2.5 | 2.5 | 2.2 | 2.3 | 3.0 | 2.4 | 2.2 |
| Arginine | 6.5 | 6.6 | 7.2 | 6.1 | 8.4 | 7.5 | 10.0 |

[a] Data of Wieser, H., Seilmeier, W., and Belitz, H., *Z. Lebensm. Unters. Forsch*, 170, 17, 1980. With permission.

aggregate in the presence of low levels of salt.[43] This aggregation tendency is probably associated with interprotein hydrophobic interactions.[43] Prolamins are usually extracted with aqueous alcohols such as 70% ethanol or 55% propan-2-ol following removal of salt-soluble proteins with saline solution. In some cases, a disulfide reducing agent such as 2-mercaptoethanol and/or acetic acid have been added,[25,29,44] which increases the yield of prolamins at the expense of the glutelins. Other factors such as temperature,[17,25] alcohol concentration,[17,45] and defatting[46,47] can also have large effects upon prolamin extractability.

Wieser et al.[23] and Ewart[24] extracted the various protein fractions from a wide variety of cereals by traditional "Osborne" procedures. The results of Wieser et al.[23] indicated that maize had the highest concentration of prolamins (48%), while wheat, sorghum, and barley showed levels ranging from 25 to 34%. Levels of rye prolamins were slightly lower (21%), while oats gave a value of 14%. Rice had a very low proportion of prolamins at 2%. Results by Ewart[24] showed similar trends, although values for prolamins were much lower for each corresponding cereal.

Orth and Bushuk[27] studied the "Osborne" solubility distribution of flour (endosperm) proteins in 26 cultivars of wheat. The prolamin (gliadin) fraction extracted in 70% ethanol accounted for 30 to 40% of the total flour nitrogen. However Miflin et al.[18] were able to extract over 60% of wheat nitrogen as prolamins using 50% propan-1-ol and 1% acetic acid at 60°C. Shewry et al.[48] increased the extraction of barley prolamins (hordein) to approximately 50% by direct extraction of ground seed with 50% propan-2-ol containing 2% 2-mercaptoethanol. With maize, Landry and Moureaux[35] found that, in contrast to barley and wheat, the addition of 2-mercaptoethanol to the 60% ethanol extraction solution resulted in only very small increases in prolamin extractability. However Sodek and Wilson[49] were able to extract an additional 18% of total maize seed nitrogen as prolamin by adding 2-mercaptoethanol, although their initial extraction with 55% 2-propanol extracted less (average = 38%) than the 49% value obtained by Landry and Moureaux.[35]

Cereal prolamins generally consist of a large number of protein components as evidenced by electrophoretic studies.[50,51] Gel filtration studies indicate that prolamins of wheat, rye, and barley have structural similarities. Prolamin extracts from all three cereals can be separated into four molecular fractions of over 100,000, 60 to 85,000, 30 to 50,000, and less than 20,000.[28,51-53] In all three cereals the 30 to 50,000 fraction is predominant and consists of single polypeptide chains that are probably stabilized by intrachain disulfide bonds. In contrast, maize prolamins consist mainly of disulfide bonded polypeptide dimers with approximate molecular weights of 45,000 and polypeptide monomers with molecular weights of 19,000 and 22,000.[54] Millet and sorghum prolamins appear to have similar structural properties to maize.[55] Oat prolamins have been separated into three fractions by gel filtration.[34,56] The major fraction, which accounted for approximately 85% of the total protein nitrogen, had a molecular weight of approximately 22,000. Because of the very low concentration of prolamins in rice, few studies on these proteins have been carried out. However, Juliano and Boulter[32] found that reduced and alkylated rice prolamins had a single major subunit with a molecular weight of 23,000 using SDS-PAGE.

Amino acid compositions of cereal prolamins are characterized by high levels of glutamic acid (mainly as glutamine) and proline and low levels of basic amino acids. Wieser et al.[23] have carried out the most complete comparative study of the amino acid compositions of cereal protein fractions. Their results (Table 3), which are similar to other published data on individual cereals, show that prolamins of wheat, rye, and barley are similar. All three cereal prolamin fractions have very high proportions of glutamic acid (plus glutamine), which averages approximately 35 mol% and of proline (16.9 to 23.4%). Maize and sorghum differ from wheat, rye, and barley in that levels of glutamic acid (plus glutamine) and proline are lower, while levels of alanine and leucine are higher. Maize and sorghum also have the lowest levels of lysine. Amino acid compositions of oat and rice prolanine differ somewhat from the other groups. Oats has glutamic acid (plus glutamine) levels approaching those of wheat, barley, and rye, but less proline, while rice is similar to maize and sorghum in glutamic acid (plus glutamine), but contains less proline.

*3. Glutelins*

As with prolamins, glutelins are generally considered to act primarily as storage proteins,

**Table 3**
**AMINO ACID COMPOSITION OF CEREAL PROLAMINES**
**(mol %)[a]**

|  | Wheat | Rye | Barley | Maize | Sorghum | Oats | Rice |
|---|---|---|---|---|---|---|---|
| Aspartic acid | 2.7 | 2.4 | 1.7 | 5.0 | 7.0 | 2.3 | 7.5 |
| Threonine | 2.2 | 2.5 | 2.0 | 3.0 | 3.7 | 2.2 | 2.8 |
| Serine | 5.5 | 6.2 | 4.3 | 6.4 | 6.1 | 3.6 | 7.0 |
| Glutamic acid | 37.7 | 36.0 | 35.9 | 19.7 | 22.5 | 34.6 | 20.0 |
| Proline | 16.9 | 18.7 | 23.4 | 10.3 | 8.1 | 10.4 | 5.2 |
| Glycine | 3.0 | 4.6 | 2.3 | 2.6 | 1.5 | 2.7 | 5.9 |
| Alanine | 2.9 | 3.1 | 2.4 | 13.8 | 13.9 | 5.6 | 9.3 |
| 1/2 Cystine | 2.2 | 2.2 | 1.9 | 1.0 | 1.1 | 3.4 | 0.8 |
| Valine | 4.0 | 4.1 | 3.7 | 3.8 | 6.1 | 7.2 | 6.5 |
| Methionine | 1.1 | 1.0 | 0.9 | 1.1 | 1.8 | 2.1 | 0.5 |
| Isoleucine | 3.9 | 2.9 | 3.4 | 3.7 | 5.0 | 3.1 | 4.4 |
| Leucine | 7.0 | 5.9 | 6.2 | 18.7 | 13.8 | 10.8 | 12.1 |
| Tyrosine | 2.0 | 1.7 | 2.3 | 3.6 | 2.2 | 1.7 | 6.3 |
| Phenylalanine | 4.7 | 4.6 | 5.9 | 5.0 | 5.1 | 5.4 | 4.9 |
| Lysine | 0.8 | 1.0 | 0.5 | trace | 0.0 | 1.0 | 0.5 |
| Histidine | 1.7 | 1.2 | 1.2 | 1.1 | 1.3 | 1.1 | 1.5 |
| Arginine | 1.7 | 1.9 | 2.0 | 1.2 | 0.8 | 2.8 | 4.8 |

[a] Data of Wieser, H., Seilmeier, W., and Belitz, H., *Z. Lebensm. Unters. Forsch*, 170, 17, 1980. With permission.

although some may act as structural proteins.[51] Glutelins are generally present in protein bodies although, with the exception or rice, their concentrations are considerably less than prolamins.[41] In rice, glutelins are the major storage proteins and are the major components of the protein bodies.[57,58]

Glutelins are normally considered to be the protein fractions remaining after extraction with saline water and aqueous alcohol. Using this definition, Wieser et al.[23] found that rye had the lowest concentration of glutelins (24.5%), while rice had the highest (77.3%). Values for wheat, barley, oats, sorghum, and maize ranged from 41.8 to 53.9%. Higher values were reported for rye, oats, and maize by Ewart.[24]

Amino acid compositions of cereal glutelins from the comparative study of Wieser et al.[23] are given in Table 4. All cereal glutelins had high concentrations of glutamic acid (plus glutamine), although values were much higher in wheat compared to the other cereals. High levels of proline were also evident in wheat and barley, while much lower levels were present in oats and rice. In general, the glutelins had amino acid compositions intermediate to the prolamins and salt-solubles.

In most cereals, the glutelins appear to consist of a diverse group of proteins with high apparent molecular weights usually ascribed to interchain disulfide bonding[40] and to hydrophobic bonding.[43] In wheat,[59] barley,[51,60] and maize,[35] glutelins have been separated into four distinct fractions by sequential extraction with aqueous alcohol in the presence of 2-mercaptoethanol, pH 10.0 borate buffers with 2-mercaptoethanol and detergent (SDS, lauryl sodium sulfate, etc.) with 2-mercaptoethanol. The three extractable fractions had properties resembling prolamins, intermediate to prolamins and salt-soluble proteins, respectively. In oats the major glutelin fraction appears to be, in fact, mainly globulin in nature.[61]

## III. STORAGE PROTEIN HYDROLYSIS DURING GERMINATION

When cereals germinate, the endosperm reserve proteins are degraded into their constituent

## Table 4
### AMINO ACID COMPOSITION OF CEREAL GLUTELINS (mol %)[a]

|  | Wheat | Rye | Barley | Maize | Sorghum | Oats | Rice |
|---|---|---|---|---|---|---|---|
| Aspartic acid | 3.8 | 7.2 | 5.0 | 5.6 | 7.8 | 9.5 | 9.7 |
| Threonine | 3.5 | 4.6 | 4.1 | 4.1 | 5.0 | 4.1 | 4.1 |
| Serine | 6.8 | 6.5 | 6.3 | 5.7 | 5.5 | 6.2 | 6.3 |
| Glutamic acid | 30.7 | 20.1 | 24.7 | 16.3 | 17.2 | 19.4 | 15.9 |
| Proline | 12.2 | 9.6 | 14.5 | 11.7 | 8.6 | 5.6 | 5.2 |
| Glycine | 8.1 | 9.4 | 6.5 | 7.0 | 7.0 | 8.1 | 7.6 |
| Alanine | 4.5 | 7.4 | 5.7 | 9.6 | 10.3 | 6.7 | 8.1 |
| 1/2 Cystine | 1.4 | 0.8 | 0.5 | 1.8 | 1.7 | 1.2 | 1.2 |
| Valine | 4.5 | 5.6 | 6.8 | 5.7 | 6.2 | 5.9 | 6.6 |
| Methionine | 1.3 | 1.6 | 1.3 | 2.8 | 1.6 | 1.3 | 2.5 |
| Isoleucine | 3.3 | 3.5 | 3.8 | 3.2 | 3.9 | 4.4 | 4.3 |
| Leucine | 7.0 | 7.5 | 7.7 | 11.1 | 9.3 | 8.0 | 8.6 |
| Tyrosine | 2.5 | 2.3 | 1.7 | 2.9 | 3.0 | 2.9 | 3.7 |
| Phenylalanine | 3.7 | 3.9 | 4.1 | 3.4 | 3.8 | 4.9 | 4.4 |
| Lysine | 2.1 | 4.1 | 2.8 | 2.4 | 3.2 | 3.3 | 3.4 |
| Histidine | 1.8 | 2.0 | 2.0 | 3.4 | 2.3 | 2.4 | 2.1 |
| Arginine | 2.8 | 3.9 | 2.5 | 3.3 | 3.6 | 6.1 | 6.3 |

[a] Data of Wieser, H., Seilmeier, W., and Belitz, H., *Z. Lebensm. Unters. Forsch*, 170, 17, 1980. With permission.

peptides and amino acids by increasing levels of proteolytic enzymes. These hydrolysis products are then transferred to the scutellum where they are either directly or indirectly utilized by the growing embryo. Although in general terms, this overall process is similar in all cereals; differences in the patterns of storage protein hydrolysis are clearly evident. Cereals that are related genetically, and thus tend to have similar storage protein and proteolytic enzyme systems, tend to show much closer similarities in their hydrolysis patterns during germination, including similar rates of hydrolysis and distributions of hydrolysis products compared to those cereals that are less closely related genetically.

### A. Barley, Wheat, and Rye

Barley, wheat, and rye are all members of the Gramineae family, subfamily Festucoideae, and tribe Hordeae. When these cereals germinate, there is usually a lag period of approximately 2 days after which a rapid phase of endosperm storage protein hydrolysis occurs.[62-66] After approximately 6 days, under more or less optimum conditions, the major portion of the endosperm storage proteins have been degraded.[62-64,67-69] The rapid increase in storage protein hydrolysis after 2 days is associated with large increases in proteolytic activity.[64,65,67] This latter topic will be discussed in detail in the next section.

Several studies have been published concerning changes in the protein fractions of barley, wheat, and rye during germination. Jahn-Deesbach and Schipper[66] studied changes in protein solubility distributions in whole seeds of wheat, barley, and rye during 84 hr of germination. The water-soluble albumin fraction showed large increases for all three cereals, while the salt-soluble globulin fraction showed little change during the germination period. The increase in water-soluble proteins was probably due to increases in albumin-like proteins in the embryo. The major endosperm storage proteins, consisting of alcohol-soluble prolamins and alkali-soluble glutelins, showed large decreases after approximately 48 hr. The rate of decrease in the prolamin fraction was greater than that for the glutelins, although in absolute terms the amount of glutelin degraded was similar to the amount of prolamin degraded in both wheat and barley. In rye only a small portion of the glutelins were degraded.

Folkes and Yemm[62] studied changes in the endosperm proteins of barley during germination. After germinating seed for various periods up to 10 days, endosperm was removed and extracted sequentially with salt solution, hot (80°C) 70% ethanol and ethanolic sodium hydroxide. All protein fractions showed rapid decreases after a 2-day lag period. Of the major storage protein fractions, the hot 70% ethanol soluble prolamins and the ethanolic hydroxide insoluble proteins tended to be degraded first, followed by the ethanolic hydroxide soluble glutelins. These results were in contrast to earlier studies of Bishop,[70] who found that (hordenins) glutelins were degraded faster than (hordeins) prolamins. However, these differences may have been due to differences in extraction conditions and nomenclature of the various fractions.[62]

Dell'Aquila et al.[69] studied changes in the protein fractions of three types of wheat including *Triticum aestivum, T. turgidum,* and *T. monococcum* during germination. Whole seeds were germinated for periods up to 6 days and then extracted sequentially with 0.5 $M$ NaCl (albumins plus globulin), 70% ethanol (prolamins), and 0.1 $M$ NaOH (glutelins). Both of the major storage protein fractions (prolamins and glutelins) were degraded rapidly during germination at similar rates. After 6 days, both fractions were almost completely depleted. In contrast the salt-soluble fractions, which probably included a high proportion of embryo proteins, showed slight increases during germination. Similar results were reported by Preston et al.[64] In this study, seeds of *T. aestivum* were germinated up to 5 days. Following removal of the embryo, the endosperm was extracted sequentially with salt solution and dilute acetic acid. Both the dilute acetic acid soluble proteins (prolamins and soluble glutelins) and the acid-insoluble glutelins showed rapid and similar rates of hydrolysis between 2 and 5 days of germination, while the salt-soluble fractions increased. Using a different approach, Hwang and Bushuk[65] studied gel filtration profiles of flour proteins from sprouted wheats, which were extracted in a strongly disassociating solvent (aqueous acetic acid-urea-cetyltrimethyl ammonium bromide). During an 8-day germination period, the very high-molecular-weight glutelin fraction and the major prolamin fractions disappeared and were replaced by a low-molecular-weight fraction (peptides and amino acids).

In both wheat and barley, the major products of storage protein hydrolysis present in the endosperm during germination have been shown to be small peptides and amino acids.[62,64,65,71,72] In contrast, there does not appear to be any build-up of intermediate hydrolysis products such as large polypeptides. For example, on the basis of amide content, Folkes and Yemm[62] concluded that there was no change in the properties of the various protein solubility fractions during germination of barley. Electrophoretic patterns of protein fractions in wheat also showed minimal changes during germination.[63,65,69] On the basis of these studies, it was concluded that during germination, individual proteins, when subjected to proteolytic attack, are degraded very rapidly to peptides and amino acids. However, it was shown that some differences in the rates of hydrolysis of individual storage proteins were evident. In particular, fractions of higher electrophoretic mobility were less affected during germination than those of lower electrophoretic mobility.[63]

## B. Oats

Oats is a member of the Gramineae family, subfamily Festucoideae, and tribe Hordeae. As in barley and wheat, the hydrolysis of oat storage proteins during germination shows an initial lag period of approximately 2 days, followed by a rapid depletion of the major endosperm reserves.[56,66,73,74] This rapid phase of storage protein hydrolysis has been associated with large increases in the levels of proteolytic activity.[73] After approximately 5 to 7 days, the bulk of the endosperm storage protein has been hydrolyzed.[56]

Jahn-Deesbach and Schipper,[66] using an "Osborne"-like fractionation procedure, found that during 84 hr of germination there was a decrease in the globulin, prolamin and glutelin fractions in whole oats and an increase in the water-soluble fraction. However, the decreases in the former protein fractions during this period were less than that obtained with wheat, barley, and rye. Similar comparative results for the prolamin fractions of wheat, barley, rye, and oats during germination were obtained by Dalby and Tsai.[75]

Kim et al.[56] studied changes in the protein solubility fractions and electrophoretic properties of oat endosperm during germination. The nonprotein nitrogen fraction (amino acids and peptides) showed large increases up to 3 days after germination, and then decreased. The salt-soluble ablumin and globulin fractions and the 45% ethanol-soluble prolamin fraction showed decreases during germination as did the predominant 0.1 $M$ acetic acid-soluble and insoluble glutelins. However, the hydrolysis rates varied for each of these fractions as well as for the various proteins within groups as determined by electrophoresis. In contrast to wheat and rye, electrophoresis revealed the formation of new bands in all fractions except the glutelins.

### C. Rice

Rice is a member of the Gramineae family, subfamily Orzoideae, and tribe Oryzeae. During germination of rice, several studies have shown decreases in endosperm protein nitrogen and concommitant increases in the levels of proteolytic enzymes.[78-80] Horiguchi and Kitagishi[77,78] showed that the decrease in endosperm protein nitrogen was due almost entirely to the hydrolysis of the glutelin fraction, which forms the major protein reserves in rice. The minor globulin and prolamin fractions showed little change during 10 days of germination, while the albumin and nonprotein nitrogen fractions increased. Evidence with protein synthesis inhibitors indicated that the increase in the albumin fraction was not due to protein synthesis, but were the hydrolysis products from the degradation of glutelins. Thus the intermediate hydrolysis products of rice, i.e., albumin-like polypeptides, appear to be hydrolyzed less rapidly than in wheat, barley, and oats during germination. Also the germination time required to deplete the major storage protein fraction in rice is longer than that required by the above mentioned cereals. After 10 days of germination, approximately one third of the endosperm glutelin was still undegraded.[78-80]

### D. Maize and Sorghum

Maize and sorghum are members of the Gramineae family, subfamily Panicoideae, and tribe Andropoganeae. During germination of these cereals, there is an initial lag period of 1 to 3 days after which the endosperm storage proteins are rapidly degraded.[81-93] This rapid degradation appears to occur in response to increasing levels of proteolytic enzymes.[87-90,94]

In sorghum and maize, the germination period required for the hydrolysis of the endosperm storage proteins appears to be longer than that required in barley, wheat, rye, and oats. Wu and Wall[81] found that in sorghum, approximately 9 days of germination was required for the hydrolysis of the prolamin (kafirin) and cross-linked prolamin fractions. In the same time period, little change had occurred in the glutelin fraction. Dure[82] showed that in germinating maize there was a steady loss of endosperm nitrogen from 3 to 10 days. By day 10, maize endosperm had lost 71% of its original nitrogen content. After 8 days of germination, Harvey and Oaks[87] found that approximately 65% of maize endosperm nitrogen had been depleted. This loss of nitrogen corresponded to losses in the major storage protein fractions (prolamins and glutelins) that were almost completely degraded. In contrast, the albumin plus globulin plus dializable-nitrogen fraction showed increased levels in the endosperm up to 80 hr after inhibitor, and then decreased. The degradation of the major storage protein fractions between 3 and 8 days coincided with the appearance of a protease with an acid pH optimum.

Moureaux[91] studied changes in the endosperm protein fractions of germinating maize over a 7-day period. No major changes in total nitrogen, protein nitrogen, or nonprotein nitrogen occurred during the first 2 days. Following this lag period, there was a rapid loss of the major endosperm proteins (prolamins and glutelins) over a 5-day period. Other minor endosperm fractions consisting mainly of basic proteins (albumins, globulins, and basic glutelins) also showed decreases.

The major products of storage protein hydrolysis present in the endosperm of maize during germination have been shown to be amino acids and small peptides.[84,86,89] Electrophoretic studies of maize prolamin patterns during germination suggest that these proteins are either hydrolyzed directly to small peptides and amino acids or that larger polypeptide hydrolysis products are degraded very rapidly. Studies by Oaks and Beevers[84] with incubated endosperms suggests that small peptides rather than amino acids are the major hydrolysis products.

## IV. ROLE OF PROTEOLYTIC ENZYMES IN STORAGE PROTEIN HYDROLYSIS

In the previous section, the changes that occurred in storage proteins upon germination were discussed. In this section, the causative agents responsible for storage protein breakdown, namely proteolytic enzymes, will be examined. Of necessity some of this discussion will be speculative as it is impossible at this stage to unequivocably ascertain cause and effect in such a complex system as the germinating cereal seed.

It should be noted that it is not the intention of the authors to go into detail concerning the overall properties of cereal proteolytic enzymes since this topic is covered in Chapters 5 and 6, Volume I. Nomenclature and methodology are discussed in Chapters 1 and 2, Volume I, respectively.

Rather, emphasis will be focused on a discussion of the specific enzymes that appear to have a major effect on storage protein hydrolysis and their synergistic mode of action in various cereals. Criteria of importance are the specificity of the enzyme for its natural substrate, a location in proximity to the substrate, and a high level of enzymic activity. In general, the enzymes of germinating cereals that appear to meet these requirements are endopeptidases with an acid pH-optimum and serine-carboxypeptidases. Both types of enzyme are normally present in high levels in the endosperm during germination and have specificities that make them capable of degrading cereal storage proteins or their intermediate hydrolysis products. Other enzyme systems such as basic endopeptidases, "BAPAases", dipeptidases and various other proteases either lack the specificity to hydrolyze storage proteins or are present at low levels in the endosperm, suggesting that they do not have major impact. Their role in transport mechanisms[95,96] as well as in as yet to be defined roles indicate, however, that we should not underestimate their importance in the overall germination process.

In general, it has been found that rates of storage protein hydrolysis in germinating cereals are closely related to the general levels of proteolytic activity. During the early stages of germination (up to 2 days), proteolytic activity and the rate of storage protein hydrolysis is relatively low, after which both increase rapidly. For simplicity, therefore, we will arbitrarily consider proteolytic enzymes in terms of such early and late germination, knowing that there is no clearly defined distinction between the two stages. As mentioned in the previous section, cereals that are related genetically have similar proteolytic enzymes and rates of hydrolysis and will, therefore, be discussed together.

### A. Barley, Wheat, and Rye

The proteolytic enzymes, which are present in the resting seed of wheat, are known to include carboxypeptidases,[97] aminopeptidases,[98] endopeptidases,[99,100] and various peptidases.[100,101]

Carboxypeptidase activity is predominantly in the endosperm, with smaller amounts residing in the outer tissues.[97] At least two major and one minor component are present as indicated by ion-exchange chromatography. An endopeptidase is also present,[99] of which approximately one quarter is present in the endosperm and the remainder in other tissues. These enzymes form during the later stages of grain development, and of necessity some mechanism such as enzyme compartmentalization or presence of inhibitors must be present to prevent storage protein breakdown. For example, studies have shown that proteolytic enzymes present in the aleurone tissue of ungerminated wheat are particulate in nature and could be released into solution by mechanical rupture of cells.[102,103] The low moisture and insolubility of the substrate could also be factors. The following evidence from studies carried out during early germination points to the presence of inhibitors of carboxypeptidase activity.

1. Carboxypeptidase inhibitors have been detected in wheat.[104]
2. No new carboxypeptidases are formed upon germination as indicated by ion-exchange chromatography, but the amounts of the individual component enzymes increase.[97]
3. Levels of the enzymes were insensitive to levels of gibberelic acid, abscissic acid, and inhibitors of protein and RNA synthesis.[105] On the other hand, proteolytic inhibitors do not appear to be involved in control of the endopeptidase enzymes.

Thus, in contrast to exopeptidase activity, increases in endopeptidase activities are minimal during early germination.[64,105] However, new endopeptidase enzymes are formed during later germination.[106] (after 2 days). These increases in endopeptidase activity appear to be a result of hormonally induced *de novo* synthesis.[105]

The role of the exo- and endopeptidase enzymes of wheat during early germination is not clear-cut, as very little degradation of the storage protein is observed during this period.[64] A possible role for these enzymes is the degradation of soluble proteins to amino acids in order to facilitate early protein synthesis in the aleurone and embryo. More subtle roles have been suggested by Mayer and Shain[107] and include a regulatory control of germination. For example, they may be involved in releasing or activating bound, particulate, or masked enzymes during early germination.

At the more advanced stages of germination, the role of specific proteolytic enzymes becomes more clear-cut. Degradation of storage proteins in the endosperm tissue proceeds, simultaneous with a four- to sixfold increase in endopeptidase activity.[64,108] Amino acids are the predominant products of protein degradation; peptide levels do not increase significantly during germination.[63,64] This suggests that the initial products of endopeptidase attack must be degraded extremely rapidly. The likely enzymes responsible for this are the carboxypeptidases, which are present in abundance and have broad specificities with the known ability to hydrolyze wheat gluten proteins[109] (Chapter 5, Volume I). Finally, most of the above studies have been carried out on half-seeds and inferences based on the anatomical location of the enzymes strongly indicates that this is the mode of breakdown in the endosperm. Whether an identical mechanism serves to break down the storage proteins present in the protein bodies of the aleurone grains remains to be established.

In barley, the proteolytic enzymes appear to be similar to those in wheat.[110] Several excellent reviews are present on the types and specificities of barley proteolytic enzymes, including the one found in this book (Chapter 5, Volume I). There appears to be at least 5 carboxypeptidases[71,111,112] and endopeptidases ranging in number from 2 to 16[113,114] that are responsible for the hydrolysis of hordein storage proteins. In the mature seed there is an inhibitor of endopeptidase activity, which could function to prevent possible autolysis.[115] The major increase in endopeptidase activity during germination, however, occurs by *de novo* synthesis.[116] Although endosperm modification occurs progressively from the embryoside of the kernel, there is still much controversy concerning the relative roles of the tissues responsible for hydrolytic enzyme formation during early germination. As with wheat, there is a rapid mobilization of endosperm reserve proteins between the 3rd to the 7th day of germination[71] time when endopeptidase activity is increasing rapidly. Concomitant with decreases in the storage proteins are the simultaneous formation of amino acids and, in particular, proline, which is one of the most abundant amino acids in hordein.[71] Although circumspect, numerous autolysis studies on malt also indicate that amino acids and TCA-soluble amino nitrogen are the major products liberated.[115,117-119] Carboxypeptidases are implicated as being largely responsible for such changes. As very little changes are observed in the hordein polypeptide electrophoresis patterns during early germination,[120] this indicates that large polypeptides, once solubilized, must be rapidly degraded to amino acids and small peptides. An added role may be ascribed to carboxypeptidases in barley by recent research, which indicates that at least one of these enzymes is able to effect the solubilization of endosperm cell walls.[112,121,122] As such, the enzyme might be extremely important in facilitating the interaction of hydrolytic enzymes with their substrates.

Although very little research has been carried out on the mobilization of protein storage reserves in rye, triticale, and oats, the enzyme systems appear to be similar to wheat and barley.[110,123,124] One might, therefore, presume that similar mechanisms are operative. With oats the rate of increase in proteolytic activity has been correlated with the rate of disappearance of protein nitrogen.[73] During germination, the hydrolysis of the five protein groups — albumins, globulins, prolamins, acetic acid-soluble, and residual glutelins — proceeds at different rates.[56] Interesting is the finding using polyacrylamide electrophoresis that, except for glutelin, newly formed protein bands are produced, which may indicate that the carboxypeptidases are less effective in degradation of these particular breakdown products.

## B. Maize and Sorghum

Recent research has indicated that the proteolytic enzymes present in cereals such as maize, sorghum, and rice, which tend to be grown in tropical and semitropical climates, have a much higher ratio of endopeptidase to carboxypeptidase activity than temperate zone cereals such as wheat and barley.[110] Because of this, it can be expected that mobilization of the endosperm reserves could be quite different.

In maize, there are at least three endopeptidases in the endosperm with pH optima of 3.8, 5.4, and 7.5 and carboxypeptidase enzymes.[87,90,125,126] As in barley,[115] there is also an endopeptidase inhibitor in maize that disappears upon germination.[127] Results by Harvey and Oaks[87] indicate that at least part of the endopeptidase enzyme is synthesized *de novo* in the maize endosperm during germination. Sufficient gibberelic acid is present in the endosperm to affect this synthesis.[128]

A number of the studies mentioned above have found that endopeptidase activity, which increases in the maize endosperm during germination, coincides with decreases in total nitrogen and protein breakdown, but the timing of these events is somewhat uncertain. For example, Harvey and Oaks[87] have found that zein and glutelin degradation in the endosperm began after 20 hr of germination, and the loss in total protein took place between 3 to 8 days, coinciding with increases in endopeptidase activity. Similarly Feller et al.[90] found that the endopeptidase and carboxypeptidase activity increased rapidly from day 2 onward, reaching a plateau between day 4 and day 6. This was accompanied by a simultaneous decrease in endosperm nitrogen. Moureaux,[91] on the other hand, found that within the first 2 days of germination, disaggregation of a portion of the glutelins into albumins and globulins can be observed. This is followed, between 2 and $2^{1}/_{2}$ days, by extensive breakdown of storage protein fractions coincident with the rate of appearance of proteolytic activity that has a maximum at $3^{1}/_{2}$ days, and thereafter decreases. The suggestion was made that the level of protease formed during this time period is sufficient to hydrolyze the storage proteins.

Evidence of a limited role for the carboxypeptidase enzymes relative to the endopeptidase enzymes in maize storage protein catabolism comes from studies using hemoglobin as substrate. Extracts from maize, sorghum, and rice liberated a great deal less carboxy-terminal amino acids than that from wheat or barley.[110] This does not preclude some role for the carboxypeptidase system, and it has been suggested by Feller et al.[90] that this enzyme system could work synergistically with the endopeptidase to increase the rate of storage protein hydrolysis. It has also been established that hydrolysis of zein during germination is accompanied by the formation of free amino acids, and in particular, phenylalanine and tyrosine.[89] This suggests the definite participation of enzymes with exopeptidase activity. Decreases in the two main zein subunits were also followed during germination by electrophoresis. No evidence of intermediate polypeptides with electrophoretically different mobilities could be detected.[89]

Although little research has been carried out on the mechanism of mobilization of storage reserves in sorghum, it appears to be different in some respects from those cereals mentioned above. In the resting seed, there is an endopeptidase, which is closely associated with protein bodies and spherosomes,[129] and a carboxypeptidase the location of which has not been ascertained.[110] Upon germination, there is also the formation of endopeptidases[94,130,131] and increases in carboxypeptidase activity.[110] Interestingly, the formation of the endopeptidase appears to be due to *de novo* synthesis, but it is not under gibberellic acid control.[94] Morphological examination by scanning electron microscopy of the breakdown of the endosperm storage reserves indicated that the matrix proteins were substantially altered prior to degradation of the protein bodies. Evidence was also obtained that, although the major degradation of the storage protein was external, some internal digestion had also occurred. This raised the unresolved question of whether the origin of the proteolytic enzymes were from the scutellum or aleurone layers, or whether they were simply activated *in situ*.

## C. Rice

As with many of the cereals described thus far, not enough research has been carried out to unequivocably link specific proteolytic enzymes in rice with storage protein hydrolysis, although, in general, increases in enzyme levels parallel decreases in endosperm storage nitrogen. In the mature seed there is at least one endopeptidase[77,79,133] and a protease inhibitor,[134] as well as carboxypeptidase activity located in the endosperm[80,135] adjacent to or in protein bodies, known to contain the glutelins. As germination proceeds, the level of this carboxypeptidase decreases, while two new carboxypeptidases are formed.[80,135] In addition, the protease inhibitors decrease[134] and endopeptidase activity increases.[78] At least part of the increase in endopeptidase is due to *de novo* synthesis,[78] as proteolytic activity was repressed by protein synthesis inhibitors. These inhibitors were less effective if applied once germination proceeded, and a novel mechanism for the mobilization of the endosperm reserve proteins was proposed. It was suggested that the hydrolytic enzymes, which were synthesized during germination, serve to disintegrate the endosperm compartmentation. This in turn allows the existing proteolytic enzymes to come in contact with the storage proteins and degrade them.

## REFERENCES

1. **Kyle, D. J. and Styles, E. D.**, Development of aleurone and sub-aleurone layers in maize, *Planta*, 137, 185, 1977.
2. **Khoo, U. and Wolf, M. T.**, Origin and development of protein granules in maize endosperm, *Am. J. Bot.*, 57, 1042, 1970.
3. **Burr, B. and Burr, F. A.**, Zein synthesis in maize endosperm by polyribosomes attached to protein bodies, *Proc. Natl. Acad. Sci. U.S.A.*, 73, 515, 1976.

4. **Larkins, B. A. and Hurkman, W. J.**, Synthesis and deposition of zein in protein bodies of maize endosperm, *Plant Physiol.*, 62, 256, 1978.
5. **Miflin, B. J., Burgess, S. R., and Shewry, P. R.**, The development of protein bodies in the storage tissues of seeds: subcellular separations of homogenates of barley, maize and wheat endosperms and of pea cotyledons, *J. Exp. Bot.*, 32, 199, 1981.
6. **Bechtel, D. B. and Juliano, B. O.**, Formation of protein bodies in the starch endosperm of rice (*Oryza sativa* L.): a reinvestigation, *Ann. Bot.*, 45, 503, 1980.
7. **Cameron-Mills, V. and von Wettstein, J.**, Protein body formation in the developing barley endosperm, *Carlsberg Res. Commun.*, 45, 577, 1980.
8. **Campbell, W. P., Lee, J. W., O'Brien, T. P., and Smart, M. G.**, Endosperm morphology and protein body development in developing wheat grain, *Aust. J. Plant Physiol.*, 8, 5, 1981.
9. **Buttrose, M. S.**, Ultrastructure of the developing wheat endosperm, *Aust. J. Biol. Sci.*, 16, 305, 1963.
10. **Bechtel, D. B., Gaines, R. L., and Pomeranz, Y.**, Early stages in wheat endosperm formation and protein body initiation, *Ann. Bot.*, 50, 507, 1982.
11. **Burgess, S. R., Turner, R. H., Shewry, P. R., and Miflin, B. J.**, The structure of normal and high-lysine barley grains, *J. Exp. Bot.*, 33, 1, 1982.
12. **Pernollet, J.-C. and Mossé, J.**, Characérisation des corpuscles porteiques de l'albumen des caryopses de céréales par microanalyse elementaire associée à la microscopic electronique a balayage, *C. R. Hebd. Seanc. Acad. Sci.* (Paris), 290D, 267, 1980.
13. **Miflin, B. J. and Burgess, S. R.**, Protein bodies from developing seeds of barley, maize, wheat and peas: the effects of protease treatment, *J. Exp. Bot.*, 33, 251, 1982.
14. **Osborne, T. B.**, The proteins of barley, *J. Am. Chem. Soc.*, 17, 539, 1895.
15. **Osborne, T. B. and Harris, I. F.**, Nitrogen in protein bodies, *J. Am. Chem. Soc.*, 22, 323, 1903.
16. **Osborne, T. B.**, The proteins of the wheat kernel, *Carnegie Inst. Washington Publ.*, 84, 1, 1907.
17. **Mossé, J.**, Monographie sur une protéine du mais: la zéine, *Ann. Physiol. Vég.*, 3, 105, 1961.
18. **Miflin, B. J., Byers, M., Field, J. M., and Faulks, J. A.**, The isolation and characterisation of proteins extracted from whole milled seed, gluten and developing protein bodies of wheat, *Ann. Technol. Agric.*, 29, 133, 1980.
19. **Richardson, M.**, The protease inhibitors of plants and microorganisms, *Phytochemistry*, 16, 159, 1977.
20. **Mikola, J. and Kirshi, M.**, Differences between endospermal and embryonal trypsin inhibitors in barley, wheat and rye, *Acta. Chem. Scand.*, 26, 787, 1972.
21. **Deponte, R., Parlamenti, R., Petrucci, T., Silano, V., and Tomasi, M.**, Albumin alpha-amylase inhibitor families from wheat flour, *Cereal Chem.*, 53, 805, 1976.
22. **Kirsi, M. and Mikola, J.**, Occurrence of proteolytic inhibitors in various tissues of barley, *Planta*, 96, 281, 1971.
23. **Wieser, H., Seilmeier, W., and Belitz, H.**, Vergleichende untersuchungen über partielle aminosäuresequenzen von prolaminen und glutelinen verschiedener getreidearten. I. Proteinfraktioneirung nach Osborne, *Z. Lebensm. Unters. Forsch*, 170, 17, 1980.
24. **Ewart, J. A. D.**, Fractional extraction of cereal flour proteins, *J. Sci. Food Agric.*, 19, 241, 1968.
25. **Byers, M., Miflin, B. J., and Smith, S. J.**, A quantitative comparison of the extraction of protein fractions from wheat grain by different solvents, and of the polypeptide and amino acid composition of the alcohol-soluble proteins, *J. Sci. Food Agric.*, 34, 447, 1983.
26. **Huebner, F. R.**, Wheat flour proteins and their functionality in baking, *Bakers Dig.*, October, 25, 1977.
27. **Orth, R. A. and Bushuk, W.**, A comparative study of the proteins of wheats of diverse baking qualities, *Cereal Chem.*, 49, 268, 1972.
28. **Preston, K. R. and Woodbury, W.**, Amino acid composition and subunit structure of rye gliadin proteins fractionated by gel filtration, *Cereal Chem.*, 52, 719, 1975.
29. **Landry, J.**, Extraction séquencée des protéines du grain d'Orge., *C. R. Acad. Sci. Paris*, 288, 907, 1979.
30. **Rhodes, A. P. and Gill, A. A.**, Fractionation and amino acid analysis of the salt-soluble protein fractions of normal and high-lysine barleys, *J. Sci. Food Agric.*, 31, 467, 1980.
31. **Bishop, L. R.**, The composition and quantitative estimation of barley proteins. I., *J. Inst. Brew.*, 34, 101, 1928.
32. **Juliano, B. O. and Boulter, D.**, Extraction and composition of rice endosperm glutelin, *Phytochemistry*, 15, 1601, 1976.
33. **Villareal, R. M. and Juliano, B. O.**, Properties of glutelin from mature and developing rice grain, *Phytochemistry*, 17, 177, 1978.
34. **Kim, S. I., Charbonnier, L., and Mossé, J.**, Heterogeneity of avenin, the oat prolamine; fractionation, molecular weight and amino acid composition, *Biochim. Biophys. Acta*, 537, 22, 1978.
35. **Landry, J. and Moureaux, T.**, Hétérogeneite des glutélines du grain de mais: extraction sélective et composition en acids ámines des trois fractions isoleés., *Bull. Soc. Chim. Biol.*, 52, 1021, 1970.
36. **Johari, R. P., Mehta, S. L., and Naik, M. S.**, Changes in protein fractions and leucine-[$^{14}$C] incorporation during sorghum grain development, *Phytochemistry*, 16, 311, 1977.

37. **Danielson, C. E.,** Seed globulins of the Gramineae and Leguminosae, *Biochem. J.,* 44, 387, 1949.
38. **Chen, C. H. and Bushuk, W.,** Nature of proteins in Triticale and its parental species. III. A comparison of their electrophoretic patterns, *Can. J. Plant Sci.,* 50, 25, 1970.
39. **Djurtoft, R.,** Salt soluble proteins of barley, *Dansk Vibenskabs Forlag A/S Kabenhorn,* 1961.
40. **Meredith, O. B. and Wren, J. J.,** Determination of molecular-weight distribution in wheat flour proteins by extraction and gel filtration in a dissociating medium, *Cereal Chem.,* 43, 169, 1966.
41. **Pernollet, J. C.,** Protein bodies of seeds: ultrastructure, biochemistry, biosynthesis and degradation, *Phytochemistry,* 17, 1473, 1978.
42. **Shuey, W. C. and Gilles, K. A.,** A note on the effect of water-flour ratio on flour protein extracted by employing a paint shaker, *Cereal Chem.,* 50, 161, 1973.
43. **Preston, K. R.,** Effects of neutral salts upon wheat gluten protein properties. I. Relationship between the hydrophobic properties of gluten proteins and their extractability and turbidity in neutral salts, *Cereal Chem.,* 58, 317, 1981.
44. **Shewry, P. R., Field, J. M., Kirkman, M. A., Faulks, A. J., and Miflin, B. J.,** The extraction solubility and characterization of two groups of barley storage polypeptides, *J. Exp. Bot.,* 31, 121, 1980.
45. **Lauriére, M., Charbonnier, L., and Mossé, J.,** Nature et fractionnement des protéines de l'Orge extraites par l'éthanol, l'isopropanol et le *n*-propanol à des titres différents, *Biochimie,* 58, 1235, 1976.
46. **Chung, K. H. and Pomeranz, Y.,** Acid soluble proteins of wheat flours. I. Effect of delipidation on protein extraction, *Cereal Chem.,* 55, 230, 1978.
47. **Charbonnier, L., Tercé-Laforgue, T., and Mossé, J.,** Rye prolamines: extractability, separation and characterization, *J. Agric. Food Chem.,* 29, 968, 1981.
48. **Shewry, P. R., Ellis, J. R. S., Pratt, H. M., and Miflin, B. J.,** A comparison of methods for the extraction and separation of hordein fractions from 29 barley varieties, *J. Sci. Food Agric.,* 29, 433, 1978.
49. **Sodek, L. and Wilson, C. M.,** Amino acid composition of proteins isolated from normal, opaque-2 and floury-2 corn endosperms by a modified Osborne procedure, *J. Agric. Food Chem.,* 19, 1144, 1971.
50. **Richetti, P. G., Gianazza, E., Viotti, A., and Soau, C.,** Hetergeneity of storage proteins in maize, *Planta,* 1136, 115, 1977.
51. **Shewry, P. R. and Miflin, B. J.,** Characterization and synthesis of barley seed proteins, in *Seed Proteins: Biochemistry, Genetics, Nutritive Value,* Gottschalk, W. and Muller, H. P., Eds., Junk, The Hague, 1983, chap. 5.
52. **Bietz, J. A. and Wall, J. S.,** Wheat gluten subunits: molecular weights determined by sodium dodecyl sulfate-polyacrylamide gel electrophoresis, *Cereal Chem.,* 49, 416, 1972.
53. **Preston, K. R. and Woodbury, W.,** Properties of wheat gliadins separated by gel filtration, *Cereal Chem.,* 53, 180, 1976.
54. **Holder, A. A. and Ingversen, J.,** Peptide mapping of the major components in vitro synthesized barley hordein: evidence of structural homology, *Carlsberg Res. Commun.,* 43, 177, 1978.
55. **Pernollet, J.-C. and Mossé, J.,** Structure and location of legume and cereal seed storage proteins, in *Seed Proteins,* Daussant, J., Mossé, J., and Vaughan, J., Eds., Academic Press, London, 1983, chap. 7.
56. **Kim, S., Pernollet, J., and Mossé, J.,** Evolution des protéines de l'albumin et de l'ultrastructure du caryopse d'Avena sativa au cours de la germination, *Physiol. Vég.,* 17, 231, 1979.
57. **Mitsuda, H., Yasumoto, K., Murakmi, K., Kusano, T., and Kishida, J.,** Studies on the proteinaceous subcellular particles in rice endosperm: electron-microscopy and isolation, *Agric. Biol. Chem.,* 31, 293, 1967.
58. **Scherebakov, V. G., Iranova, D. I., and Fedorova, S. A.,** Structure and chemical composition of proteins in the endosperm of rice, *Soviet Plant Physiol.,* 20, 757, 1973.
59. **Ryadchikov, V. G., Zimah, V. G., Zhamkina, O. A., and Lebeder, A. V.,** A study of the glutenins and gliadins of wheaten flour (in Russian), *Prikladnaya Biokhimiya i Microbiologiya,* 17, 25, 1981.
60. **Landry, J. T., Moreaux, T., and Huet, J. C.,** Extractabilité des protéins de grain d'orge: dissolution sélective et composition en acid amines des fractions isolés., *Biologique,* 7-8, 281, 1972.
61. **Robert, L. S., Nozzolillo, C., Cudjoe, A., and Altosaar, I.,** Total solubilization of groat proteins in high protein oats (*Avena sativa* L. cv. Hinoat): evidence that glutelins are a minor component, *Can. Inst. Food Sci. Technol. J.,* 16, 196, 1983.
62. **Folkes, B. F. and Yemm, E. W.,** The respiration of barley plants. X. Respiration and the metabolism of amino-acids and proteins in germinating grains, *New Phytol.,* 57, 106, 1957.
63. **Coulson, C. B. and Sim, A. K.,** Wheat proteins. II. Changes in the protein composition of *Triticum vulgare* during the life cycle of the plant, *J. Sci. Food Agric.,* 16, 499, 1965.
64. **Preston, K. R., Dexter, J. E., and Kruger, J. E.,** Relationship of exoproteolytic and endoproteolytic activity to storage protein hydrolysis in germinating durum and hard red spring wheat, *Cereal Chem.,* 55, 877, 1978.
65. **Hwang, P. and Bushuk, W.,** Some changes in the endosperm proteins during sprouting of wheat, *Cereal Chem.,* 50, 147, 1973.

66. **Jahn-Deesbach, W. and Schipper, A.,** Gekeimten kornern von weizen, gerste, roggen und hafer, Getreide, *Mehl. Brot.*, 34, 281, 1980.
67. **Beresh, I. D.,** Proteolysis of gluten during sprouting of wheat (in Russian), *Proc. All-Union Sci. Res. Inst. Grain Grain Prod.*, 66, 111, 1969.
68. **Metivier, J. R. and Dale, J. E.,** The utilization of endosperm reserves during early growth of barley cultivars and the effect of time of application of nitrogen, *Ann. Bot.*, 41, 715, 1977.
69. **Dell'Aquila, A., Colaprico, G., Taranto, G., and Carella, G.,** Endosperm protein changes in developing and germinating *T. aestivum*, *T. turgidum* and *T. monococcum* seeds, *Cereal Res. Commun.*, 11, 107, 1983.
70. **Bishop, L. R.,** The changes undergone by the nitrogenous constituents of barley during malting. I, *J. Inst. Brew.*, 35, 323, 1929.
71. **Mikola, L. and Mikola, J.,** Mobilization of proline in the starchy endosperm of germinating barley grain, *Planta*, 149, 149, 1980.
72. **Wagner, D. and Piendle, A.,** Effect of barley variety, environment and malting technology on the amino acids of malt. II, *Brew. Dig.*, 51(4), 50, 1976.
73. **Sutcliffe, J. F. and Baset, Q. A.,** Control of hydrolysis of reserve materials in the endosperm of germinating oat (*Avena sativa* L.) grains, *Plant Sci. Lett.*, 1, 15, 1973.
74. **Wu, Y.V.,** Effect of germination on oats and oat protein, *Cereal Chem.*, 60, 418, 1983.
75. **Dalby, A. and Tsai, C. Y.,** Lysine and tryptophan increases during germination of cereal grains, *Cereal Chem.*, 53, 222, 1976.
76. **Matsushita, S.,** Studies on the nucleic acid in plants. III. Changes in the nucleic acid contents during the germination stage of the rice plant (in Japanese), *Mem. Res. Inst. Food Sci. Kyato Univ.*, 14, 30, 1958.
77. **Horiguchi, T. and Kitagishi, K.,** Studies on rice seed protease. II. Changes in protease activity and nitrogen compounds of germinating rice seed (in Japanese), *J. Sci. Soil Manure Jpn.*, 40, 225, 1969.
78. **Horiguchi, T. and Kitagishi, K.,** Protein metabolism in rice seedling. I. Effect of inhibitors of protein synthesis on the degradation of seed protein during germination, *Soil Sci. Plant Nutr.*, 22, 327, 1976.
79. **Palmiano, E. P. and Juliano, B. O.,** Biochemical changes in the rice grain during germination, *Plant Physiol.*, 49, 751, 1972.
80. **Doi, E., Komori, N., Matoba, T., and Morita, Y.,** Some properties of carboxypeptidases in germinating rice seeds and rice leaves, *Agric. Biol. Chem.*, 44, 77, 1980.
81. **Wu, Y. V. and Wall, J. S.,** Lysine content of protein increased by germination of normal and high-lysine sorghums, *J. Agric. Food Chem.*, 28, 455, 1980.
82. **Dure, G. S.,** Gross nutritional contributions of maize endosperm and scutellum to germination growth of maize axis, *Plant Physiol.*, 35, 919, 1960.
83. **Ingle, J., Beevers, L., and Hageman, R. H.,** Metabolic changes associated with the germination of corn. I. Changes in weight and metabolites and their redistribution in the embryo axis, scutellum and endosperm, *Plant Physiol.*, 35, 735, 1964.
84. **Oaks, A. and Beevers, H.,** The requirement for organic nitrogen in *Zea mays* embryos, *Plant Physiol.*, 39, 37, 1964.
85. **Oaks, A.,** The regulation of nitrogen loss from maize endosperm, *Can. J. Bot.*, 43, 1077, 1965.
86. **Ingle, J. and Hageman, R. H.,** Metabolic changes associated with the germination of corn. III. Effects of gibberellic acid on endosperm metabolism, *Plant Physiol.*, 40, 672, 1965.
87. **Harvey, B. M. R. and Oaks, A.,** The hydrolysis of endosperm protein in *Zea mays*, *Plant Physiol.*, 53, 453, 1974.
88. **Harvey, B. M. R. and Oaks, A.,** The role of gibberellic acid on the hydrolysis of endosperm reserves in *Zea mays*, *Planta*, 121, 67, 1974.
89. **Fujimaki, M., Abe, M., and Arai, S.,** Degradation of zein during germination of corn, *Agric. Biol. Chem.*, 41, 887, 1977.
90. **Feller, U., Soong, T. T., and Hagemen, R. H.,** Patterns of proteolytic enzyme activities in different tissues of germinating corn (*Zea mays* L.), *Planta*, 140, 155, 1978.
91. **Moureaux, T.,** Protein breakdown and protease properties of germinating maize endosperm, *Phytochemistry*, 18, 1113, 1979.
92. **Bose, B. and Srivastava, H. S.,** Proteolytic activity and nitrogen transfer in maize seeds during inhibition, *Biol. Plant.*, 22, 414, 1980.
93. **Bose, B., Srivastava, H. S., and Mathur, S. N.,** Effect of some nitrogenous salts on nitrogen transfer and protease activity in germinating *Zea mays* L. seeds, *Biol. Plant.*, 24, 89, 1982.
94. **Kohler, D. E.,** Control of protease activity during sorghum germination, *Proc. Plant Growth Regul. Work Group*, 7, 128, 1980.
95. **Sopanen, T., Burston, D., and Matthews, D. M.,** Uptake of small peptides by the scutellum of germinating barley, *FEBS Lett.*, 79, 4, 1977.
96. **Mikola, J. and Kohlemainen, L.,** Localization and activity of various peptidases in germinating barley, *Planta*, 104, 167, 1972.

97. **Kruger, J. E. and Preston, K.,** The distribution of carboxypeptidases in anatomical tissues of developing and germinating wheat kernels, *Cereal Chem.,* 54, 167, 1977.
98. **Kruger, J. E. and Preston, K.,** Changes in aminopeptidases of wheat kernels during growth and maturation, *Cereal Chem.,* 55, 360, 1978.
99. **Preston, K. and Kruger, J.,** Location and activity of proteolytic enzymes in developing wheat kernels, *Can. J. Plant Sci.,* 56, 217, 1976.
100. **Kruger, J. E.,** Changes in the levels of proteolytic enzymes from hard red spring wheat during growth and maturation, *Cereal Chem.,* 50, 122, 1973.
101. **Pett, L. B.,** Studies on the distribution of enzymes in dormant and germinating wheat seeds. I. Dipeptidase and protease, 29, 1898, 1935.
102. **Rowsell, E. V. and Goad, L. J.,** Some effects of Gibberellic acid on wheat endosperm, *Proc. Biochem. Soc.,* 90, 11P, 1964.
103. **Gibson, R. A. and Paleg, L. G.,** Lysosomal nature of hormonally induced enzymes in wheat aleurone cells, *Biochem. J.,* 128, 367, 1972.
104. **Preston, K. R. and Kruger, J. E.,** The nature and role of proteolytic enzymes during early germination, *Cereal Res. Commun.,* 4, 213, 1976.
105. **Preston, K. R. and Kruger, J. E.,** Physiological control of exo- and endoproteolytic activities in germinating wheat and their relationship to storage protein hydrolysis, *Plant Physiol.,* 64, 450, 1979.
106. **Preston, K. R.,** Note on separation and partial purification of wheat proteases by affinity chromatography, *Cereal Chem.,* 55, 793, 1978.
107. **Mayer, A. M. and Shain, Y.,** Control of seed germination, *Annu. Rev. Plant Physiol.,* 25, 167, 1974.
108. **Fleming, J. R., Johnson, J. A., and Miller, B. S.,** Effect of malting procedure and wheat storage conditions on alpha-amylase and protease activities, *Cereal Chem.,* 37, 363, 1960.
109. **Preston, K. R. and Kruger, J. E.,** Purification and properties of two proteolytic enzymes with carboxypeptidase activity in germinated wheat, *Plant Physiol.,* 58, 516, 1976.
110. **Winspear, M. J., Preston, K. R., Rustagi, V., and Oaks, A.,** Comparison of peptide hydrolase levels in cereals, *Plant Physiol.,* 1984, in press.
111. **Ray, L. E.,** Large scale isolation and partial characterization of some carboxypeptidases from malted barley, *Carlsberg Res. Commun.,* 41, 169, 1976.
112. **Baxter, E. D.,** Purification and properties of malt carboxypeptidases attacking hordein, *J. Inst. Brew.,* 84, 271, 1978.
113. **Sundblom, N. O. and Mikola, J.,** On the nature of the proteinases secreted by the aleurone layer of barley grain, *Physiol. Plant.,* 27, 281, 1972.
114. **Burger, W. C. and Schroeder, R. L.,** A sensitive method for detecting endopeptidases in electrofocused thin-layer gels, *Anal. Biochem.,* 71, 384, 1976.
115. **Mikola, J. and Enari, T. M.,** Changes in the contents of proteolytic inhibitors during autolysis and malting potential, *J. Inst. Brew.,* 86, 216, 1980.
116. **Jacobsen, J. V. and Varner, J. E.,** Gibberellic acid-induced synthesis of protease by isolated aleurone layers of barley, *Plant Physiol.,* 42, 1596, 1967.
117. **Jones, M. and Pierce, J. S.,** Malt peptidase activity, *J. Inst. Brew.,* 73, 347, 1967.
118. **Sopanen, T., Takkinen, J., Mikola, J., and Enari, T. M.,** Rate-limiting enzymes in the liberation of amino acids in mashing, *J. Inst. Brew.,* 86, 211, 1980.
119. **Maendyl, W. D. and Piendyl, A.,** Effect of barley variety, environment and malting technology on the amino acids of malt. II., *Brew. Dig.,* April, 50, 1976.
120. **Shewry, P. R., Faulks, A. J., Parmar, S., and Miflin, B. J.,** Hordein polypeptide pattern in relation to malting quality and the varietal identification of malted barley brain, *J. Inst. Brew.,* 86, 138, 1980.
121. **Bamforth, C. W., Martin, H. L., and Wainright, T.,** A role for carboxypeptidase in the solubilization of barley β-glucan, *J. Inst. Brew.,* 85, 334, 1979.
122. **Martin, H. L. and Bamforth, C. W.,** The relationship between β-glucan solubilase, barley autolysis and malting potential, *J. Inst. Brew.,* 86, 216, 1980.
123. **Madl, R. L. and Tsen, C. C.,** Proteolytic activity of Triticale, *Cereal Chem.,* 50, 215, 1973.
124. **Nowak, R. L. and Mierzwinska, T.,** Activity of proteolytic enzymes in rye seeds of different ages, *Z. Pflanzenphysiol,* 86, 15, 1978.
125. **Abe, M. Arai, S., and Fujimaki, M.,** Purification and characterization of a protease occuring in endosperm of germinating corn, *Agric. Biol. Chem.,* 41, 893, 1977.
126. **Abe, M., Arai, S., and Fujimaki, M.,** Substrate specificity of a sulfhydryl protease purified from germinating corn, *Agric. Biol. Chem.,* 42, 1813, 1978.
127. **Abe, M., Arai, S., Kato, H., and Fujimaki, M.,** Thiol-protease inhibitors occurring in endosperm of corn, *Agric. Biol. Chem.,* 44, 685, 1980.
128. **Oaks, A., Winspear, M. J., and Misya, S.,** Hydrolysis of endosperm protein in *Zea mays* ($W_{64A}$ × $W_{18E}$), *3rd Int. Symp. Pre-Harvest Sprouting Cereals,* Kruger, J. E. and LaBerge, D. E., Eds., Westview Press, Boulder, Colo., 1983, 204.

129. **Adams, C. A. and Novellie, L.,** Acid hydrolases and autolytic properties of protein bodies and spherosomes isolated from ungerminated seeds of *Sorghum bicolor* (Linn.) Moench., *Plant Physiol.*, 55, 7, 1975.
130. **Garg, G. K. and Virupaksha, T. K.,** Acid protease from germinated sorghum. I. Purification and characterization of the enzyme, *Eur. J. Biochem.*, 17, 4, 1970.
131. **Garg, G. K. and Virupaksha, T. K.,** Acid protease from germinated sorghum. II. Substrate specificity with synthetic peptides and ribonuclease, *Eur. J. Biochem.*, 17, 13, 1970.
132. **Glennie, C. W., Harris, J., and Liebenberg, N. V. D. W.,** Endosperm modification in germinating sorghum grain, *Cereal Chem.*, 60, 27, 1983.
133. **Ozaki, K. and Horiguchi, T.,** Studies on rice germ protease. I., *Nippon Dojo-Hiryogaku Zasshi*, 36, 95, 1965.
134. **Horiguchi, T. and Kitagishi, K.,** Studies on rice seed protease. V. Protease inhibitor in rice seed, *Plant Cell Physiol.*, 12, 907, 1971.
135. **Doe, E., Komori, N., Matoba, T., and Morita, Y.,** Purification and some properties of a carboxypeptidase in rice bran, *Agric. Biol. Chem.*, 44, 85, 1980.

Chapter 2

## ROLE OF PROTEOLYTIC ENZYMES IN THE MOBILIZATION OF PROTEIN RESERVES IN THE GERMINATING DICOT SEED

### Karl A. Wilson

### TABLE OF CONTENTS

| | | |
|---|---|---|
| I. | Introduction | 20 |
| II. | Experimental Approaches to the Study of Storage Protein Mobilization | 20 |
| | A. Time Course Studies of Proteolytic Enzymes of Seeds | 20 |
| | B. Characterization of Structural Changes in Storage Proteins During Mobilization | 21 |
| III. | The Storage Proteins of Dicotyledonous Plants | 22 |
| | A. Protein Bodies | 23 |
| | B. The Legumin and Vicilin Type Storage Globulins | 23 |
| | C. Lectins | 25 |
| | D. Proteinase Inhibitors | 26 |
| | E. Other Low-Molecular-Weight Storage Albumins | 27 |
| IV. | Proteolytic Enzymes of Dicots | 27 |
| | A. Classification of Proteolytic Enzymes | 27 |
| | B. Endopeptidases of Dicots | 28 |
| |     1. Assays for Endopeptidases | 28 |
| |     2. Properties of Dicot Seed Endopeptidases | 29 |
| | C. Exopeptidases of Dicots | 30 |
| |     1. Acid Carboxypeptidases | 30 |
| |     2. Arylamidases | 31 |
| |     3. Alkaline Peptidases | 31 |
| | D. Regulation of Proteolytic Enzymes by Endogenous Protein Inhibitors | 32 |
| V. | Examples of Storage Protein Mobilization During Germination and Seedling Development | 33 |
| | A. Proteolysis of Pumpkin Storage Proteins | 33 |
| | B. Proteolysis of Legume Storage Globulins | 35 |
| | C. Proteolysis of Legume Storage Proteins Other Than Globulins — Lectins and Trypsin Inhibitors | 38 |
| VI. | The Generalized Pathway for Storage Protein Mobilization During Germination | 39 |
| Acknowledgment | | 40 |
| References | | 40 |

## I. INTRODUCTION

The seed of the spermatophytes (the seed plants) is an adaptation to the rigors of terrestrial life. The seed represents a dormant form of the embryonic sporophyte plant that can withstand the conditions of drought, frost, etc. and can serve as a dispersal agent for the plant species. When presented with suitable environmental conditions of moisture, temperature, etc. the seed may break its quiescence and undergo the process of germination. Water is imbibed from the environment, rehydrating the seed tissues. Metabolic processes such as respiration rapidly increase in rate. The process of germination is finally completed by the emergence of the embryo, usually the radicle first, through the various layers investing the seed. Under continuing suitable conditions, the seedling continues to grow and develop, eventually producing a functional, photosynthetic plant.

During germination and initial stages of seedling development, the plant is dependent upon the nutrient reserves laid down in the seed during its development on the mother plant. These reserves must supply energy, preformed carbon chains, organically fixed nitrogen, and minerals to the developing seedling until it has developed a fully functional root system and the photosynthetic capability characteristic of an autotrophic plant. The most commonly found reserve or storage molecules are carbohydrates (including starch and oligosaccharides such as raffinose and stachyose), proteins, lipids (especially triglycerides), and phytins (which serve as a reserve of metal ions and phosphorus). The relative proportions of each of the storage forms varies widely between different plant families and species. In the mature seed of dicotyledonous plants, the reserves are generally located in the cotyledons of the embryo, or the endosperm, or both.

During germination and seedling growth, the reserves are mobilized, that is, the storage molecules are hydrolyzed to their constituent units, e.g., monosaccharides, amino acids, fatty acids, etc. These are then utilized by the seedling for its energy metabolism and biosynthesis. In this chapter we will examine our current knowledge of the enzymology of storage protein mobilization in the dicot seed. The reader is also directed to previous reviews, which have examined various aspects of this problem.[1-7]

## II. EXPERIMENTAL APPROACHES TO THE STUDY OF STORAGE PROTEIN MOBILIZATION

Two basically different approaches to the study of dicot storage protein mobilization have appeared in the literature: (1) characterization of the temporal changes in the proteolytic enzymes present in quiescent and germinating seeds and in the developing seedling; and (2) characterization of the structural changes occurring in the storage protein during germination and early seedling development. Ideally, the study of storage protein mobilization in a given species of plant would involve both of these experimental approaches. Unfortunately this has seldom been the case in practice.

### A. Time Course Studies of Proteolytic Enzymes of Seeds

The majority of studies involving protein mobilization in seeds have addressed the phenomenon by examining the enzymes that may be involved. In addition to examining the physical, chemical, and enzymatic properties of purified or partially purified enzymes, time course studies are often carried out. The assumption underlying this approach is that those proteolytic activities that are present or increase in level coincident with the disappearance of the storage proteins are those enzymes responsible for the degradation.

The basic experimental design used is to follow the levels of a number of different types of proteolytic activities in the seed at different stages of development. These stages generally range from the dry quiescent seed to the developing seedling where the majority of the storage proteins have been depleted from the storage organs. A key point to note here is that the experimenter actually measures an activity hydrolyzing a given substrate under a defined set of assay conditions, rather than an enzyme protein per se. The choice of substrate is critical in this type of approach. This is especially true in light of the observation that in a number of systems the enzymes catalyzing the initial step or steps of storage protein mobilization are quite specific for the native storage protein (see Sections IV.B, V.A, and V.B). Therefore, if only inappropriate animal proteins such as casein or hemoglobin are used as substrates, key endopeptidases may go undetected.

A final note of caution in interpreting time course data for both endo- and exopeptidase activities is required. While these studies will indicate whether a proteolytic enzyme occurs in the storage organ contemporaneously with the storage protein mobilization, it will not in any sense prove that said enzyme is actually involved. Like all eukaryotic cells, those of the dicots are highly compartmentalized. For an enzyme to interact with a substrate, it must at least be present in the same subcellular compartment. Most of the published time course studies have used extracts derived from whole tissue, which presumably contain proteins derived from the cytosol, ruptured protein bodies, vacuoles, and other organelles. The presence of an activity at the time of storage protein degradation does not prove it is indeed involved, even if such an activity can be demonstrated in vitro.

The author does not wish to leave the reader with the impression that the difficulties noted above invalidate the time course type study as a useful approach to the question of storage protein mobilization. Rather, such studies serve as a useful basis for further experimentation. Coupled with studies of the subcellular compartmentalization, it can yield useful data on which enzymes may indeed be involved at the various stages of mobilization. If appropriate care is taken in the choice of substrate and assay conditions, this information may extend to the initial proteolytic events inflicted on the storage proteins.

### B. Characterization of Structural Changes in Storage Proteins During Mobilization

Study of the proteolytic enzymes present during mobilization of reserves gives us little, if any, direct indication of what is actually happening to the storage proteins at the same time. For this information, we must directly examine the storage proteins themselves. A number of techniques have been used over the years for this purpose, with sophistication and sensitivity increasing as new techniques have appeared and have been applied to proteins in general.

A common observation made is an increase in the electrophoretic mobility of the storage protein(s) in polyacrylamide gels under mildly alkaline (pH 8 to 9) conditions. This has been noted in germinating peanut (*Arachis hypogaea*),[8] pea (*Pisum sativum*),[9] common bean (*Phaseolus vulgaris*),[10,11] soybean (*Glycine max*),[12,13] chickpea (*Cicer arietinum*),[14] and common vetch (*Vicia sativa*).[16] This is often interpreted as being due to an increase in the negative charge on the storage protein molecule, most likely resulting from deamidation of asparagine or glutamine residues.[8,12] Significant decreases in the amide content of vetch and chickpea legumins (17 and 31% in 7.5 and 6 days, respectively),[15,16] vetch vicilin (26% in 7.5 days),[15] and peanut total seed extract (29% in 8 days)[8] have been noted in support of this hypothesis.

However, the increased electrophoretic mobility seen with the proteins from germinated seeds could also result from partial proteolysis. Separation by polyacrylamide gel electrophoresis is based upon a combination of molecular charge, molecular mass, and molecular shape.[17] Loss of a portion of the storage protein molecule by partial proteolysis would reduce the molecular size and could increase the charge to molecular mass ratio. Both of these changes would result in an increase in electrophoretic mobility. Loss of a portion of the molecule rich (relative to the remainder of the molecule) in asparaginyl or glutaminyl residues would reduce the amide content of the protein, whether based upon weight of protein or molar quantity, without deamidation per se occurring. Further studies are thus necessary to establish the true significance of the mobility change noted in these plant species. In a number of other plant species, physicochemical studies of the storage proteins have indicated that, at least in the early stages of germination, limited proteolysis of the storage proteins such as that hypothesized above does indeed occur.

Polyacrylamide disc gel electrophoresis has proved of invaluable use in a system where a number of discrete intermediary forms appear during the mobilization of a low-molecular-weight protein, the mung bean trypsin inhibitor (see Section V.C). However, in general, the legumin and vicilin type storage proteins migrate as relatively broad diffuse bands in polyacrylamide gel electrophoresis, presumably due to their intrinsic macro- and micro-heterogeneity (see Section III).

The use of polyacrylamide disc gel electrophoresis in the presence of sodium dodecyl sulfate (SDS-PAGE)[18,19] has proved especially useful in studying the changes in a number of systems, especially those with legumin-like storage proteins.[20-24] This technique separates polypeptides on the basis of their chain length, in general without respect to the intrinsic charge of the polypeptide. It thus yields a direct indication of polypeptide molecular weight. It may be carried out with or without prior reduction of any disulfide bonds present in the sample. This feature is potentially of use when studying storage proteins that contain two or more polypeptide chains linked by disulfides, such as native legumin subunits. To date, two-dimensional gel techniques, such as that of O'Farrell,[25] have not been applied to the problem of storage protein degradation. They have, however, been used extensively in characterizing the legumins from ungerminated seeds.[20,26-31]

In either one- or two-dimensional gel techniques, one would hopefully be able to observe the disappearance of the original storage protein polypeptide(s) as mobilization proceeds. At the same time, any discrete intermediates resulting from partial proteolysis will appear as new bands or spots on the electropherogram. A necessary condition for the useful interpretation of such data is the ability to unambiguously identify the newly appearing species as degradation products of the original storage protein. This is not a trivial matter when it is considered that while the degradation of the storage protein is proceeding, a large number of other proteins are typically being synthesized. This is especially true in those cases where the storage organ takes on a new function, such as the cotyledons of many seeds that become functional photosynthetic organs after germination.

Such assignments may be made if samples of purified native and partially proteolyzed storage proteins are available as standards. Except for the native protein this is, however, seldom the case. The advent of new techniques such as immunoblotting[32-34] should greatly enhance our ability to identify the breakdown products. In this technique, the separated proteins or polypeptides are electrophoretically transferred from the electrophoresis gel to a blotting matrix, typically of nitrocellulose or a diazotized derivative of paper, where they are immobilized. The "blot" may be stained for protein in the conventional manner, or it may be reacted with antibodies to allow the visualization of specific proteins and other species with similar or identical antigenic determinants. Alternatively, the protein of interest and its derivatives may be partially purified by immunoaffinity chromatography[26] prior to electrophoresis. The resulting electropherogram should ideally in this case contain only peptides derived from the desired storage protein.

## III. THE STORAGE PROTEINS OF THE DICOTYLEDONOUS PLANTS

Osborne[35] in 1924 proposed the division of the plant proteins into four groups on the basis of their solubility properties:

1. Albumins — soluble in water at neutral or slightly acidic pH

2. Globulins — insoluble in water, but soluble in salt solutions
3. Glutelins — insoluble in water or salt solutions, but soluble in strongly acidic or basic solutions
4. Prolamins — soluble in ethanol solutions, but insoluble in water

The glutelins and prolamins are the major storage proteins in the monocotyledonous plants, but are absent or present only at low levels in the dicots. In contrast, the globulins, and occasionally the albumins, are the major storage proteins of the dicots.[1,35-41] The legumin and vicilin type storage proteins fall into the former class, while most of the lectins and proteinase inhibitors behave as albumins.

## A. Protein Bodies

In most seeds, the bulk of the storage protein appears to be localized in the protein bodies, also called the aleurone grains.[1,40,41] These subcellular organelles are found throughout the embryo of the dicot, especially in the cotyledons, but also in the axis[42-44] and the endosperm if this tissue is present as a storage organ in the mature seed.[45,46] The structure, composition, and changes in the protein bodies during germination have been the subject of a number of reviews,[1,36,40] and will be only briefly covered here.

The protein bodies appear roughly spherical in shape, with their size varying considerably both between species of plants, and indeed within the same cell of a given species. In the dicots they range between 1 and 22 μm in diameter, with an average diameter of a few microns. Protein bodies are bounded by a single membrane. The internal architecture of the protein body differs among different plant species. In some plants, especially the legumes, there appears to be a relatively homogeneous or granular structure with no apparent inclusions.[47-50]

In the seeds of other plants, one or two types of inclusions, crystalloids and globoids, may be found imbedded in a relatively homogeneous matrix within the protein body. As a general rule, nonlegume species have protein bodies with either globoids, or globoids and crystalloids. Peanut differs from other legumes in containing some protein bodies containing globoids, and some without.[51] The crystalloids appear to be partly crystalline deposits of the major storage globulin(s) of the seed, e.g., edestin in hemp,[52] or the globulin of the castor bean.[45,46] The globoids are made up primarily (60 to 80% by dry weight) of the $K^+$, $Mg^{2+}$, $Ca^{2+}$ salts of phytic acid (myoinositol hexaphosphate). In addition to the phytate, the globoids may also contain enzymes involved in the catabolism of the phytate, such as the phosphatase found in cotton globoids.[53]

## B. The Legumin and Vicilin Type Storage Globulins

Because of the economic importance of the legumes, much of the work on seed storage protein biochemistry has focused on this plant family.[37,41,54-56] Two major classes of storage globulins have been identified in the legumes, the legumins and the vicilins. The relative proportion of legumin and vicilin in the seed varies widely between different plant species.[38] Both have been shown to be localized in the protein bodies.[11,40,57] The relationships between storage proteins from different plants has at times been obscured by the practice of assigning different names to globulins isolated from specific plant species. Thus, while the peanut and soybean both clearly contain proteins closely related to the vicilin of the broad bean (*Vicia faba*), the proteins from these two species are termed α-conarachin and β-conglycinin, respectively. Similar names, arachin and glycinin, have been applied to the legumins of peanuts and soybeans, respectively.[58]

The legumins typically have sedimentation coefficients of approximately 11S (range 10S to 15S, depending upon the plant species and conditions), with molecular weights in the range of 300,000 to 400,000. Their amino acid compositions typically exhibit high contents

## Table 1
## AMINO ACID COMPOSITIONS OF STORAGE PROTEINS[a]

|      | Legumin[b] | Vicilin[b] | BBSTI[b,c] | KSTI[b,d] | Lectin[b] | Albumin[e] |
|------|------------|------------|------------|-----------|-----------|------------|
| Asx  | 13.4       | 13.7       | 15.5       | 14.4      | 13.5      | 4.4        |
| Thr  | 3.9        | 2.4        | 2.8        | 3.9       | 7.0       | 1.4        |
| Ser  | 5.4        | 7.1        | 12.7       | 6.1       | 10.6      | 11.8       |
| Glx  | 16.6       | 22.3       | 9.8        | 9.9       | 7.2       | 30.0       |
| Pro  | 6.4        | 4.8        | 8.4        | 5.5       | 6.5       | 2.5        |
| Gly  | 7.2        | 4.6        | 0          | 8.8       | 5.6       | 8.4        |
| Ala  | 5.4        | 5.2        | 5.6        | 4.4       | 8.5       | 3.7        |
| ½Cys | 1.6        | 0          | 19.7       | 2.2       | tr[f]     | 8.5        |
| Val  | 5.9        | 4.0        | 1.4        | 7.7       | 6.3       | 4.6        |
| Met  | 1.6        | 0          | 1.4        | 1.1       | 0.8       | 1.6        |
| Ile  | 4.9        | 4.6        | 2.8        | 7.7       | 5.3       | 3.0        |
| Leu  | 7.4        | 9.2        | 2.8        | 8.3       | 8.8       | 4.2        |
| Tyr  | 3.0        | 2.4        | 2.8        | 2.2       | 2.2       | 1.8        |
| Phe  | 4.7        | 5.4        | 2.8        | 5.0       | 5.2       | 0.9        |
| Lys  | 4.5        | 6.0        | 7.0        | 5.5       | 5.6       | 3.0        |
| His  | 1.9        | 1.4        | 1.4        | 1.1       | 1.9       | 0.8        |
| Arg  | 6.1        | 6.8        | 2.8        | 5.0       | 2.5       | 9.4        |
| Trp  | ND[f]      | ND[f]      | 0          | 1.1       | 2.5       | tr[f]      |
| Ref. | 59         | 68         | 206        | 90        | 207       | 108        |

[a] All values expressed as residues/100 residues, recalculated from original data when required.
[b] From soybean, *Glycine max.*
[c] BBSTI, Bowman-Birk soybean trypsin inhibitor; calculated from amino acid sequence.
[d] KSTI, Kunitz soybean trypsin inhibitor; calculated from amino acid sequence.
[e] From castor bean (*Ricinus communis*) endosperm.
[f] Trace, tr; not determined, ND.

of glutamic acid/glutamine, and generally relatively high contents of aspartic acid/asparagine and arginine (Table 1). The protein is often referred to in the literature simply as the 11S protein or globulin. The legumin of soybean, glycinin, has been extensively studied. It has a molecular weight of approximately 320,000 to 350,000, and contains two types of subunits, which differ in their molecular weights and isoelectric points.[59] Each glycinin molecule is composed of six "acidic" and six "basic" subunits, with each acidic chain disulfide-bonded to a basic chain.[21,22] Soybean glycinin thus has the structure $(A-B)_6$, where A and B are the acidic and basic chains, respectively. The acidic subunits have molecular weights of approximately 37,000 to 45,000,[60,61] while the molecular weight of the basic subunits is approximately 20,000.[21,59]

Both the acidic and basic chains exist as multiple forms. Six different acidic chains have been identified by Moreira et al.[61] in soybean cultivar CX635-1-1-1, while Kitamura and Shibasaki[60] noted four types of acidic chains in the Raiden cultivar. Both soybean cultivars were found to contain glycinins with four different basic chain types.[22,61] Staswick et al.[21] recently demonstrated that the pairing of acidic and basic chains is nonrandom. They identified five specific disulfide-bonded acidic chain-basic chain pairs using ion exchange chromatography, electrophoresis, and isoelectric focusing. One acidic chain ($A_4$ of their nomenclature, $M_r$ 37,000) was not found to be contained in a disulfide-bonded pair.[61] Staswick et al.[21] proposed on the basis of these data that glycinin is synthesized as a 60,000 mol wt precursor, which is subsequently hydrolyzed to form the disulfide-bonded acidic chain-basic chain pair. This indeed was subsequently shown to be true.[62-64]

The legumins of other legume species are similar in general characteristics to those of soybean glycinin,[27,28,30,54-56] and appear to be synthesized via single chain precursor molecules.[24,65] A number of nonleguminous dicots have also been demonstrated to have storage proteins of the legumin type,[55] notably the edestin of hemp,[65a] acalin B of cotton (*Gossypium hirsutum*) seed,[55] and the major globulin of pumpkin (*Cucurbita moschata*) seeds.[66] Globulins with sedimentation coefficients of 11S have been demonstrated in the seeds of many other dicot species, but have not been characterized sufficiently to definitively identify them as homologous to the legumins of the Leguminosae.[54]

The second major group of storage globulins, the vicilins, are more variable in their properties than the legumins. All of the vicilins thus far described in dicots have come from the legumes, with the exception of the acalin A of cotton seeds.[55] They have molecular weights between 110,000 and 190,000, and sedimentation coefficients between 6S and 9S. They are generally termed 7S globulins. Like legumin, they typically have high contents of glutamic acid/glutamine, aspartic acid/asparagine, and arginine (Table 1). The vicilin from a given species of legume is generally heterogeneous. This heterogeneity is possibly due, at least in part, to the fact that many vicilins are glycoproteins that may exhibit heterogeneity of the carbohydrate moiety.[54] As with the legumins, the vicilins of the economically important legumes have been the most extensively studied.

The majority of the 7S globulin fraction of the soybean is composed of the vicilin β-conglycinin,[67] which aggregates to a 10S form at low ionic strength.[68] Thanh and Shibasaki[68,69] separated purified β-conglycinin into six distinct species ($B_1$ to $B_6$) by ion exchange chromatography. Examination of the β-conglycinins by isoelectric focusing and electrophoresis indicated the presence of three major subunits, α, α′, and β.[68-70] They further observed that the β-subunit could be resolved into four closely similar species on gel isoelectric focusing.[70] The molecular weights of the subunits were found to be approximately 57,000 for α and α′, and 42,000 for β. A fourth minor subunit, γ, was found to have a molecular weight similar to that of β. Each of the native β-conglycinin forms was demonstrated to be a trimer made up of a combination of one or more of the three major subunits: B1, $\alpha'\beta_2$; B2, $\alpha\beta_2$; B3, $\alpha\alpha'\beta$; B4, $\alpha_2\beta$; B5, $\alpha_2\alpha'$; and B6, $\alpha_3$.[68] A seventh form, $\beta_3$, has been described by Sykes and Gayler.[71]

The presence of multiple forms has been noted in the vicilins of other legumes, including peas and broad beans.[56] The subunit compositions of the vicilins from other legume species is, however, less clear than that of the soybean β-conglycinins. In a number of other legumes, up to eight presumed vicilin subunits have been identified, with molecular weights ranging from 57,000 to 19,000.[54,56] It is possible, and indeed probable, that most of the smaller "subunits" are in fact degradation products of larger subunits. This concept was suggested by the finding that the 24,000 and 29,500 mol wt polypeptides of the mung bean vicilin increase in prevalence after treatment of the vicilin in vitro with the mung bean vicilin peptidohydrolase.[72] Chrispeels et al.[65] have demonstrated that pea vicilin is synthesized as polypeptides of 75,000, 70,000, 50,000, and 49,000 mol wt, with the smaller vicilin polypeptides ($M_r$ 34,000, 30,000, 25,000, 18,000, 14,000, 13,000, and 12,000) arising from post-translational modification of either the 49,000 or 50,000 mol wt polypeptides, or both.

## C. Lectins

The lectins, also known as phytohemagglutinins, may also be considered as potential storage proteins in some species of plants. These proteins were initially noted due to their ability to agglutinate erythrocytes from man and a number of other animal species. The lectins may be defined as sugar-binding proteins of nonimmune origin, which are devoid of enzymatic activity towards the sugars that they bind.[74] They have proven to be especially common in the seeds of plants, especially the legumes,[75,76] where they are found in fairly large concentrations. The jackbean (*Canavalia ensiformis*) lectin concanavalin A represents 2 to 3% of the seed protein, while the soybean and common bean lectins make up 1 to 1.5%

and 10 to 20%, respectively, of the total seed protein.[77-79] During germination, most of the lectins are degraded along with the bulk of the storage proteins.[80]

The lectins of most legumes have molecular weights of approximately 110,000 to 135,000 and are tetramers of subunits of 25,000 to 35,000 mol wt. The subunits may all be identical, as in the case of concanavalin A, or several different types of subunits may be found.[76] In the latter case, a number of "isolectin" forms may be found. For example, red kidney beans (*Phaseolus vulgaris*) lectin contains two types of subunits, E and L.[75] These may be combined to form isolectins with subunit compositions of $E_4$, $E_3L$, $E_2L_2$, $E_1L_3$, and $L_4$.

The majority of plant lectins are glycoproteins, most commonly containing mannose, *N*-acetylglucosamine, glucose, and in some cases, galactose, fucose, xylose, and arabinose.[76] They also generally contain bound metal ions, typically $Mn^{2+}$ and $Ca^{2+}$, which are necessary for sugar binding.

Subcellular fractionation and microscopy studies indicate that the lectins are largely located in the protein bodies of a number of dicots.[11,45,46,57,81,82] However, one class of lectins, the β-lectins (so called because of their specificity for β-glycosides) is found in the intercellular spaces, the cell wall, and in the peripheral cytoplasm.[83]

## D. Proteinase Inhibitors

The plant proteinase inhibitors are a diverse group of generally low-molecular-weight proteins. They share in common the ability to form complexes with animal and some plant proteinases, thereby inactivating the proteinases.[4,84-86] They are present in especially high concentrations in seeds, particularly those of legumes.[87] Essentially all of the inhibitors that have been well characterized chemically have come from this plant family. The proteinase inhibitors may be conveniently divided into several families on the basis of their molecular weights, inhibitory properties, and amino acid sequences.[84] Two inhibitor families have been extensively studied in the Leguminosae, the Bowman-Birk type trypsin inhibitors, and the Kunitz type inhibitors.

The Bowman-Birk type inhibitors appear to be found in the seeds of all or nearly all legumes. They are approximately 8000 in molecular weight, although some members of the family show a strong tendency towards self-association in solution.[88] This property has often led to confusion in the literature in interpreting molecular weight data from techniques such as gel filtration. The inhibitors exist in their native, unmodified state as single polypeptide chains of 60 to 85 amino acid residues. They have a characteristic amino acid composition (Table 1) containing high levels of half-cystine (14 residues per molecule, approximately 20 residue percent), all of which is present as cystinyl residues. Relatively high levels of aspartic acid/asparagine and of serine are also found, while methionine, valine, and the aromatic amino acids are generally low. Tryptophan is generally absent.[86] Most legume species examined have been found to contain two or more distinct isoinhibitor species.[86] A number of animal and bacterial enzymes have been observed to be inhibited by Bowman-Birk type inhibitors, including mammalian trypsin, chymotrypsin, and elastase, as well as the bacterial proteinase subtilisin.[84,86]

The Kunitz type inhibitors are exemplified by the Kunitz soybean trypsin inhibitor (KSTI).[89] KSTI has a molecular weight of 21,300, consisting of a single polypeptide chain of 181 residues, including four residues of half-cystine and two methionyl residues (Table 1).[90] In recent years, it has been shown that the single autosomal gene locus coding for KSTI exists as four allelic forms, $Ti^a$, $Ti^b$, $Ti^c$, which are codominant and yield functional trypsin inhibitors, and ti, a recessive allele where no functional inhibitor is synthesized.[91,92] Soybean strains are thus possible with up to two different KSTI forms present.

Recently, Kunitz-type proteinase inhibitors have been described from a number of other legume species.[93-95] They are of similar size (160 to 200 amino acid residues; $M_r$ 18,000 to 22,000) to KSTI and clearly homologous on the basis of amino acid sequence. However, in a number of these species the inhibitors are found to consist of two polypeptide chains of 130 to 160 residues and 39 to 44 residues, respectively. The larger of these chains is in each case homologous to the amino-terminal portion of KSTI, while the smaller chain is homologous to the carboxyl-terminal region of KSTI. Odani and co-workers[96] have suggested that these inhibitors and synthesized as single-chain precursors, which are subsequently modified to the two-chain form found in the dry seed.

Subcellular fractionation studies with the mung bean (*Vigna radiata*),[97] hyacinth bean (*Lablab purpureaus*),[98] pea,[99] and common bean[11] indicate that the Bowman-Birk type inhibitors are cytosolic in localization. No data are available as to the localization of the Kunitz type inhibitors within the cell.

A number of physiological functions have been suggested for the plant proteinase inhibitors,[100,101] including (1) protection of the seed from predation, either by insects or other invertebrates, or microorganisms (see Chapter 4, Volume I), (2) regulation of proteolysis by endogenous seed proteinases, or (3) serving as storage proteins, especially of sulfur-containing amino acids. It is the latter possibility that is of particular interest to us at this point. The destruction of the seed trypsin inhibitors during germination has been observed in a number of species of legumes.[86,99,101-105] The Kunitz and Bowman-Birk type inhibitors make up 1 to 6% of the total seed protein in a number of legume species.[102,106] However, the high content of half-cystine of the Bowman-Birk type inhibitors, compared to the low content of sulfur-containing amino acids in the storage globulins (Table 1), means that these inhibitors contain a disproportionately high amount of the seed's sulfur reserves.[107] The inhibitors, especially the Bowman-Birk type, are thus probably important to the legume seed as storage proteins, even if they may also fulfill another function such as control of proteolysis (see Section IV.D) or protection from exogenous proteinases.

### E. Other Low-Molecular-Weight Storage Albumins

Three low-molecular-weight albumins have been identified as major storage proteins in the endosperm of the castor bean.[108] All have apparent molecular weights of approximately 12,000 by SDS-PAGE, with sedimentation coefficients of 2S. They constitute about 40% of the total protein of the protein body, where they are localized in the proteinaceous matrix. The amino acid composition of the mixture of the three proteins is typical of storage proteins, with high contents of glutamic acid/glutamine and arginine (Table 1). The albumins are rapidly degraded during germination, largely disappearing by 2 days after imbibition. Sharief and Li[108a] have recently sequenced the *Ricinus* albumin and have shown it to consist of two polypeptide chains, one 34 residues long and the other 61 residues long. The two chains are linked by one or more disulfide bonds. An evolutionary relationship has been suggested between the albumin and the Bowman-Birk proteinase inhibitors[108a] and barley (*Hordeum vulgare*) trypsin inhibitor.[108b] Similar proteins have been identified in the embryonic axis of the mung bean,[109] and the pea.[110]

## IV. PROTEOLYTIC ENZYMES OF DICOTS

### A. Classification of Proteolytic Enzymes

Proteolytic enzymes have historically been divided into a number of different classes, using a number of different criteria (see Chapters 1 and 2, Volume I). Hartley[110a] has proposed four major groups based upon their catalytic mechanisms: serine proteinases, sulfhydryl (now cysteine) proteinases, metalloproteinases, and acid (now aspartic) proteinases. Alternatively, these enzymes may be classified in terms of their catalytic specificity. In this case, two major classes may be identified. Endopeptidases hydrolyze internal peptide bonds, with peptides produced as products. In the case of the exopeptidases, internal peptide bonds are not hydrolyzed. Rather, they have the action of hydrolyzing the peptide bond involving the carboxyl- or amino-terminal residue of either a protein or peptide. At least one of the liberated products is thus a free amino acid. This class of proteolytic enzymes has also been termed peptidases.[5,111] This manner of classification, utilizing an endopeptidase/exopeptidase dichotomy, is most applicable when considering protein mobilization in the dicots. It must, however, be pointed out that this dichotomy is not absolute, as many endopeptidases exhibit some degree of exopeptidase-like activity.[5]

### B. Endopeptidases of Dicots
*1. Assays for Endopeptidases*

Much of the early work on proteolytic enzymes concerned animal, and specifically mammalian, enzymes. As a consequence, many of the assays used in the study of plant endopeptidases have been identical to, or variations of, those developed for the animal enzymes (see Chapter 2, Volume I). The most commonly used assays for general proteolytic activity are modifications of Anson's assay, using bovine hemoglobin as a substrate, or the method of Kunitz, with casein as substrate.[112,113] The substrate is incubated with the enzyme and the appropriate buffer, with or without a reducing agent such as dithiothreitol or 2-mercaptoethanol. After the appropriate time, the unhydrolyzed substrate is precipitated by the addition of trichloroacetic acid (TCA). The products of proteolysis, small peptides, and free amino acids are then quantitated. Originally, absorbance at 280 nm was used.[112,113] However, for greater sensitivity, a number of other techniques are in current use, including reaction with ninhydrin or trinitrobenzenesulfonate (TNBS)[114] to determine the release of TCA-soluble amino groups, or reaction with the Lowry reagent. A further variation is the use of a substrate protein to which is coupled an azo-dye, e.g., azocasein.[115] Here the hydrolysis may be followed by the appearance of TCA-soluble dye chromophore.

These types of assays have several potentially critical drawbacks in the study of the proteolytic enzymes of storage protein mobilization.[1] A very specific endopeptidase, which cleaves only one or a relatively few peptide bonds in the substrate, may escape detection. If the polypeptides produced by the action of the enzyme are very large, or are bound to much larger polypeptides (e.g., via disulfide bonds), they may be precipitated by the TCA, thus masking the activity of the enzyme. For this type of enzyme, a variation of the method of Lin et al.[116] would seem to be of use. This assay technique does not require a TCA precipitation step, but rather determines newly released amino groups directly in the substrate-enzyme incubation mixture. This is possible because of the low background that the *N*-methylated substrate protein provides.

The Anson and Kunitz type assays are also less than ideal for the detection of exopeptidases, especially if absorbance at 280 nm is used as the quantitation method. However, the use of the ninhydrin or TNBS detection techniques may, nevertheless, prove successful in detecting the liberated free amino acids produced by these enzymes.

A far greater drawback to these assays is the use of an inappropriate animal protein such as casein, hemoglobin, or gelatin as a substrate. It has become increasingly clear that a number of the endopeptidases involved in the initial stages of storage protein mobilization are relatively specific for their natural protein substrates. Such endopeptidases have been described in the pumpkin[117] and vetch (*Vicia sativa*).[118] In both instances, the enzyme has comparatively little activity against commonly used substrates such as casein or hemoglobin, but is highly active with the appropriate natural substrate. It is obvious then that the use of purified storage proteins as substrates in endopeptidase assays is of critical importance.

This has indeed been done in several recent studies. Generally storage proteins from the same plant species have been used, including unfractionated seed globulins,[119-121] and isolated legumins, vicilins, and seed albumins.[121-124]

However, gliadin from wheat has been used as a substrate for lupin (*Lupinus angustifolius*) endopeptidase,[125] while Shain and Mayer[126] used a plant pectinase as substrate for the proteolytic activity of lettuce seeds (*Lactuca sativa*). Azo-dye-stained storage proteins have also been used in analogy to azocasein.[11,99,127] Spencer and Spencer[128] have described as assay for the pumpkin endopeptidase using pumpkin seed globulin to which 1-anilino-8-naphthalenesulfonate (ANS) was noncovalently bound. Hydrolysis of the globulin is detected by the decrease in the fluorescence of the ANS as it is released from the hydrophobic environment on the native protein. However, it has not yet been determined if the coupling of the azo-dye or the binding of ANS to the storage protein alters its susceptibility to degradation by the plant endopeptidases.

Synthetic peptide, amide, and ester substrates have proven useful for the assay of many animal endopeptidases, e.g., $N$-$\alpha$-benzoyl-L-arginine ethyl ester or $N$-$\alpha$-benzoyl-D,L-arginine $p$-nitroanilide for the assay of mammalian pancreatic trypsin. However, their use as substrates for the assay of putative plant endopeptidases has led to considerable confusion in the literature. "Trypsin-like" enzymes have been claimed to be found in a number of plant species using these substrates.[126,129-131] However, these studies have failed to recognize that these substrates are also hydrolyzed by some plant peptidases, such as carboxypeptidases and arylamidases. There has thus far been no definitive demonstration of a dicot seen endopeptidase functionally or structurally related to the mammalian trypsins.

*2. Properties of Dicot Seed Endopeptidases*

The endopeptidases of dormant seeds have been much less studied than those from germinated seeds. In addition, the present limited data do not suggest the uniformity of characteristics found with the germinated seed endopeptidases. Shutov et al.[132] have described an endopeptidase (protease E) from ungerminated vetch seeds that hydrolyzes casein with a pH optimum of 5.5. It is inhibited by diisopropylfluorophosphate (DFP), but not by iodoacetate. It declines rapidly during germination. This enzyme appears similar to the endopeptidase activity observed by Murray et al.[133] in crude extracts of the radicles of pea seeds imbibed for 1 day. In the case of the pea enzyme, a pH optimum for casein hydrolysis of 5.7 was found. Phenylmethylsulfonyl fluoride (PMSF) treatment resulted in 30% loss in activity, while $N$-ethylmaleimide (NEM) caused only slight inhibition. No activation by thiols was noted. Endopeptidase activities with similar acidic pH requirements have been noted in the ungerminated seeds of a number of species of legumes, including the mung bean,[100,134] soybean,[135,136] and the common bean,[137,138] as well as a number of other dicots such as cotton.[139] An endopeptidase has been partially purified from quiescent hemp (*Cannabis sativa*) by St. Angelo et al.[124,141] It exhibits an optimum for hemoglobin degradation of pH 3.4 and at pH 4.3 with the hemp storage protein edestin. It is slightly inhibited by DFP, but not by sulfhydryl-reactive reagents such as NEM or mercurials. It has a molecular weight of approximately 20,000. Bond and Bowles[136] have described a proteolytic activity in the extracts from dry soybeans, which is optimumly active against the iodinated B-chain of insulin at pH 4.0. This is apparently an aspartic proteinase, as it is strongly inhibited by pepstatin A.

Endopeptidases with near neutral pH optima have been described in a number of ungerminated seeds. Protease I of pumpkin[117] exhibits a pH optimum of 6.8 when native pumpkin globulin is used as substrate. It is slightly inhibited by ethylenediaminetetraacetate (EDTA) and $p$-hydroxymercuribenzoate (pCMB), but is unaffected by PMSF. Its activity decines as germination proceeds. Great northern beans (*P. vulgaris*) contain, in addition to an acidic endopeptidase activity, a second enzyme active at pH 7 with casein as substrate. The two activities are separable by isoelectric precipitation of the neutral endopeptidase at pH 5. Hydrolysis of casein by the neutral endopeptidase is not affected by PMSF or $p$-chloromercuribenzenesulfonate.[142] Dry soybeans have been shown to contain an endopeptidase that degrades iodonated insulin B-chain with a pH optimum of 8.0. This enzyme is inhibited by the chelators EDTA and 1,10-phenanthroline, as well as by pCMB and $Cu^{2+}$. This suggests a metalloproteinase that also has a free thiol group required for activity.[136] A similar enzyme has been described in ungerminated jackbean (*Canavalia ensiformis*).[143]

There is some question in the literature as to whether the endopeptidases of ungerminated seeds are capable of hydrolyzing the storage globulins of ungerminated seeds. Studies of the enzymes of dry seed of vetch,[144] mung bean,[145] pea,[9] and lotus (*Nymphaea* sp.)[146] indicate that the globulins prepared from a dry seed do not serve as substrates for these enzymes. However, similar studies with hemp,[124] pea,[147] broad bean,[148] soybean,[136,149] and pumpkin[117] yield the contrary result. Shutov and Vaintraub have suggested that the use of impure globulin preparations, or globulin preparations that had been acted upon by bacterial proteinases, may explain the hydrolysis noted in some cases.[150] However, it is now apparent that in some plants, such as the pumpkin and the soybean, the dry seed globulins are indeed susceptible to proteolysis by enzymes present in the dry seed.

The endopeptidases of germinated seeds have been much more extensively characterized than those present in the quiescent dry seed, presumably because of their obvious possible involvement in protein mobilization. These proteolytic enzymes exhibit considerable uniformity among dicot species, and indeed within the seed-bearing plants in general.[9,72,111,118,123,125,151-157] They are relatively low-molecular-weight (approximately 20,000 to 40,000), sulfhydryl-dependent endopeptidases.

Their pH optima range from 3.5 to 6.5, with considerable variation noted with different substrates. They are in general relatively nonspecific in their action, in the sense that they will hydrolyze animal proteins such as casein and hemoglobin as well as, generally, both native and partially degraded plant storage proteins. They are apparently found in the protein bodies, or the vacuoles derived from the fusion of protein bodies, of germinated seeds, although this has only been established definitively in two cases.[152,158] These enzyme activities are presumed to increase during germination and seedling development due to *de novo* synthesis, as has been demonstrated in the mung bean.[159]

## C. Exopeptidases of Dicots

The plant exopeptidases (see Chapter 5, Volume I) have been divided by Mikola[111] into three groups, based upon their specificity of action and substrates, and their pH optima: the acid carboxypeptidases, the naphthylamidases, and the alkaline peptidases. While this classification was orignally based largely upon data drawn from the study of monocot species such as barley and rice, it appears to be equally applicable to dicot peptidases.

### 1. Acid Carboxypeptidases

Plant carboxypeptidases differ considerably from the well-known mammalian pancreatic carboxypeptidases in many aspects. While the latter are metalloenzymes, the plant enzymes are serine dependent, being inhibited by reagents such as DFP and PMSF, and unaffected by metal ion chelating agents such as EDTA and 1,10-phenanthroline. Further, they have distinctly acidic pH optima, generally in the range of pH 5 to 6, depending upon the substrate and conditions. In contrast to the moderate to high specificities of cleavage of the mammalian carboxypeptidases A and B,[160,161] the plant carboxypeptidases are capable of releasing at significant rates all 20 of the natural amino acyl residues when present at the carboxyl-terminus of peptides or proteins.

Carboxypeptidases have been described in a number of dicots, including cotton,[162] kidney bean,[163] pea,[133] mung bean,[120] castor bean (*Ricinus communis*),[164] peanut,[165] and soybean.[166] The enzyme appears to be present in the seeds in the protein bodies, and the vacuoles formed by the fusion of protein bodies during germination.[145,164,167] In at least two plants, peanut[165] and soybean,[166] two distinct enzymes are present that differ in their substrate specificity and/or chromatographic behavior. Carboxypeptidase activity is in general found in significant levels in the ungerminated seeds of legumes, and increases during germination.[120,133,164] In cotton there is no detectable carboxypeptidase in the dormant seed. Upon germination there is a rapid increase, to a maximum at day 4 of germination. This increase has been shown to be due to *de novo* synthesis utilizing a mRNA synthesized during embryogenesis prior to dessication of the mature seed.[168]

*2. Arylamidases*

The arylamidases are characterized by their ability to catalyze the hydrolysis of the amide bond between the carboxyl group of an amino acid and the amino group of an aromatic amine (e.g., *p*-nitroaniline or β-naphthylamine). These enzymes have been found in all dicot seeds that have been examined for their presence, including the pea,[121,133,169,170] peanut,[123,165,171,172] soybean,[173] mung bean,[120,145,167] common bean,[138] vetch,[174-176] pumpkin,[157] buckwheat,[177] and castor bean.[152] They generally have pH optima in the range of pH 6.5 to 8.5, with molecular weights of approximately 60,000 to 70,000.

The arylamidases appear, at least in the mung bean, to be cytosolic in localization.[145,167] In those systems where careful studies have been carried out, multiple enzymes have been observed ranging from three to five different types in a given species of seed.[152,169,170,172,174-176] These enzymes generally appear to fall into three groups on the basis of their substrate specificities, i.e., those hydrolyzing the arylamides of (1) arginine, (2) proline, and (3) neutral and aromatic amino acids such as leucine, alanine, phenylalanine, etc. Multiple forms within a group may be present in any particular dicot species, especially in groups (1) and (3) above. The arylamidases appear to be sulfhydryl dependent, being inhibited by NEM and pCMB and also to varying extents by 1,10-phenanthroline. A number of the arginine specific arylamidases are inhibited by DFP and/or PMSF.

In a number of cases, arylamidases have been obtained in sufficient purity to more carefully assess their substrate specificity.[157,169,171,174,176] In addition to the arylamides, they have been found to hydrolyze a number of di-, tri-, and oligopeptides, which are presumably their natural substrates in vivo. In the case of the enzymes hydrolyzing arginine arylamides, larger polypeptides such as histones, prolamin, and the B-chain of insulin are also hydrolyzed.[157,176,177] This hydrolysis has been demonstrated to take place at internal Arg-X bonds, indicating that these enzymes are capable of acting as endopeptidases. In contrast, the enzymes acting on the arylamides of neutral and aromatic amino acid appear to act as aminopeptidases since blocking of the α-amino group of the substrate renders it immune to hydrolysis.[169,174] The arylamidases generally decline or remain relatively constant in activity during germination. However, in a number of species, significant increases in activity have been seen during germination, notably in the castor bean[152] and pumpkin.[157]

*3. Alkaline Peptidases*

The plant alkaline peptidases are those enzymes with alkaline (pH 7.5 to 10) pH optima that hydrolyze simple peptide substrates, but are inactive against typical arylamidase substrates. In this group are the plant leucine aminopeptidases and dipeptidases.[111] While present in significant levels in both quiescent and germinating seeds, this group of enzymes has received relatively little attention. This may in part be due to the lack of simple, direct chromogenic assays for the enzymes.

Mikola and co-workers have extensively studied this group of enzymes in the monocot barley and the gymnosperm scots pine (*Pinus sylvestris*).[5,111] However, very little data is available on these enzymes from dicots. Ashton and Dahmen[178] partially purified two closely similar leucine aminopeptidases from germinating squash (*Cucurbita maxima*) seeds using L-leucineamide as a marker substrate. The pH optimum was found to be approximately pH 8 for both enzymes, although a minor optimum was also seen at pH 7. In addition to leucineamide, the dipeptides L-leucylglycine and L-leucyl-L-leucine, and the tripeptide L-leucylglycylglycine also served as substrates. The squash aminopeptidases appear to be metallo-enzymes, being inhibited by EDTA, and activated by $Mg^{2+}$ and $Mn^{2+}$. A specific dipeptidase has also been highly purified from germinating squash seeds.[179] It is specific for L-dipeptide hydrolysis, and is inactive towards tripeptides and aminoacylamides. The pH optimum for the hydrolysis of L-leucylglycine is 8.5 or greater. The purified enzyme is weakly activated by $Mg^{2+}$ and $Mn^{2+}$, and inhibited by $Zn^{2+}$ and $Co^{2+}$. Dipeptidase activity has also been shown to be present in germinating peanuts[165] and soybeans.[180] The dipeptidase of the squash seed increases approximately fivefold during germination,[181] while the peanut enzyme undergoes only a slight increase during germination.[165]

### D. Regulation of Proteolytic Enzymes by Endogenous Protein Inhibitors

The presence of large quantities of protein proteinase inhibitors in many dicot seeds has led to the speculation that they might be involved in the control of proteolysis in either the developing seed during storage protein deposition, the quiescent seed, or the germinating seed.[4,100] However, the evidence for such a function is at present ambiguous.

The best experimental data indicating the presence of an inhibitor of an endogenous dicot seed proteinase comes from the work of Baumgartner and Chrispeels.[103] They have demonstrated that the dry mung bean seed contains a low-molecular-weight (approximately 12,000, by gel filtration) inhibitor of the major endopeptidase of the germinating seed, vicilin peptidohydrolase. This inhibitor is distinct from the Bowman-Birk type trypsin inhibitor of the mung bean,[97,104,182] which is inactive against the vicilin peptidohydrolase.[97] The inhibitor declines rapidly during germination, with nearly half of the original level gone by the 3rd day of germination. In contrast, only approximately 15% of the trypsin inhibitor is destroyed over the same time period. Baumgartner and Chrispeels suggest that this inhibitor is not involved in the regulation of the levels of the vicilin peptidohydrolase because (1) the inhibitor decreases faster during germination than the proteinase increases, and (2) the inhibitor is cytosolic in localization, while the proteinase is found in the protein bodies. They hypothesize that the function of the inhibitor is that of protection of the cytoplasm from the proteinase in the event of accidental rupture of the protein bodies/vacuoles during storage protein degradation. The presence of this type of inhibitor may explain the results of Basha and Cherry.[123] They found that extracts from ungerminated peanuts inhibited the proteolytic activity present in extracts of seeds germinated for 10 days.

The literature does not in general support the idea that the trypsin inhibitors of seeds inhibit endogenous proteolytic enzymes. However, a number of reports have appeared proposing such an interaction. The presence of a trypsin inhibitor active against a "trypsin-like" proteinase in lettuce (*L. sativa*) has been claimed by Shain and Mayer.[126,131] They hypothesize that the destruction of this inhibitor during germination controls the production of the trypsin-like enzyme from some inactive precursor form. However, it is now apparent that their "trypsin-like" enzyme is most likely the acid carboxypeptidase of the lettuce seed, which would also be expected to hydrolyze $N$-$\alpha$-benzoyl-L-arginine ethyl ester. Their contention that the enzyme arises from a precursor form, rather than *de novo* synthesis, is based in significant part upon the failure of chloramphenicol to prevent the increase in the levels of the enzyme during germination. It is now known, however, that chloramphenicol is an inhibitor of protein synthesis on prokaryotic ribosomes, or prokaryotic-like ribosomes such as those in the mitochondria and chloroplasts of eukaryotes. Chloramphenicol is thus an inappropriate probe for protein synthesis on the ribosomes of the plant cell endoplasmic reticulum, where one would expect the enzymes involved in storage protein degradation to be synthesized.[183]

Their results do, however, suggest that there is an inhibitor present in the dry seeds, which is active against the acid carboxypeptidase of the germinated seeds. There is no evidence to suggest that this is the trypsin inhibitor. Unfortunately this has not been examined further. A nonprotein carboxypeptidase inhibitor has been described from kidney beans.[184] This inhibits mammalian carboxypeptidase A and other metalloenzymes by the complexation of the metal ions, especially $Zn^{2+}$, necessary for the activity of the enzymes. It would, however, not be expected to inhibit the typical plant carboxypeptidases, which are not metalloenzymes.

Observations more indicative of trypsin inhibitors interacting with the proteinase inhibitors of ungerminated seeds have been reported by Royer and co-workers,[185,186] and Gennis and Cantor.[187] If a crude albumin preparation from ungerminated cowpea (*Vigna unguiculata*) is passed through a resin containing immobilized bovine trypsin, the trypsin inhibitor is removed through its specific interaction with the trypsin. Such treatments increase the caseolytic activity, at pH 7, of the albumin preparations. Addition of the affinity-purified trypsin inhibitor back to the inhibitor-depleted albumin results in inhibition of the caseolytic activity, suggesting that the trypsin inhibitor is indeed responsible for the inhibition. However, this study can be criticized on the grounds that immobilized trypsin resins can also act as ion exchange resins due to the large positive charge on the trypsin at most pH values (pI = 10.1).[188] The material binding to, and subsequently released from, the trypsin resin is not necessarily only trypsin inhibitor, but may also include acidic proteins that are not active as trypsin inhibitors, but could conceivably inhibit the caseolytic activity of the albumin preparation. This objection does not, however, apply to the study by Gennis and Cantor, again in *V. unguiculata*.[187] They found that a highly purified preparation of trypsin inhibitor inhibited a protease from the seeds that was active against azocasein (bovine) at pH 8.0. The significance of these observations is unclear in view of the lack of information on the proteolytic enzyme(s) involved, their subcellular localization, and the absolute degree of purity of the inhibitor preparations.

A number of other reports have appeared suggestive of inhibitors in seeds active against endogenous proteolytic enzymes. This evidence generally consists of an increase in the total units of activity recovered following a step in a purification procedure.[121,171,179,181] Interpretation of such observations is difficult, as the influence of nonprotein inhibitory factors, such as phenolics or metal ion chelators and such as that noted in the kidney bean[184] cannot be ruled out.

## V. EXAMPLES OF STORAGE PROTEIN MOBILIZATION DURING GERMINATION AND SEEDLING DEVELOPMENT

In addition to studies on the proteolytic enzymes present in germinated and ungerminated seeds, a number of studies have examined changes in the storage proteins themselves.

### A. Proteolysis of Pumpkin Storage Proteins

The mobilization of the storage protein of the pumpkin (*Curcubita moschata* and related hybrids, Curcubitaceae) has been studied by several groups, but especially that of Hara and co-workers. The seeds of the pumpkin contain a typical legumin-like globulin as their major storage protein. The globulin contains two different types of subunits, α and β, with molecular weights of 63,000 and 56,000, respectively.[66,190,195] Each subunit is in turn made up of two polypeptide chains bound together by one or more disulfide bonds. The α-subunit is composed of an acidic γ-chain ($M_r$ 36,000) and a basic $δ_1$-chain ($M_r$ 22,000). The β-subunit is likewise composed of an acidic (γ', $M_r$ 34,000) and basic ($δ_2$, $M_r$ 22,000) chain.[66,190]

The pumpkin globulin is localized in the protein bodies of the parenchymal cells of the cotyledons.[193,194] In the ungerminated seed, these protein bodies are approximately 10 μm in diameter and consist of a small globoid, a large crystalloid, and a proteinaceous matrix.[193] The crystalloid is insoluble in water, and has been shown to consist of the legumin-like storage globulin by subfractionation of the protein bodies in nonaqueous solvent followed by SDS-PAGE.[194] Within 24 hr of the beginning of imbibition, the protein bodies swell and fuse together. As germination proceeds, the crystalloid decreases in size, with a concomitant increase in the proteinaceous matrix. By day 8 of germination, the crystalloid has essentially disappeared, with only small protein particles remaining in the now vacuolar protein body.[193]

These morphological changes in the pumpkin cotyledonary protein body correlate well with changes in the physical properties of the storage protein globulin. As germination proceeds, the water-insoluble globulin is converted into a water-soluble form, until by day 4 of germination nearly all of the globulin is water soluble.[189,192,195] When examined by SDS-PAGE in the absence of reducing agents, the solubilized globulin (designated $E_{\alpha\beta}$) was found to have an apparent molecular weight of 43,000 with a minor fraction of 30,000 mol wt. Comparison with the molecular weights of 56,000 and 63,000 for the major β- and minor α-subunits of the water-insoluble globulin (designated F) suggests a limited proteolysis results in the solubilization of the globulin.

SDS-PAGE in the presence of reducing agent demonstrates that $F_{\alpha\beta}$ consists of polypeptides of approximately 20,000 mol wt. This is consistent with a proteolysis of the acidic chains (γ and γ′) of F, removing a fragment or fragments equivalent to 16,000 to 14,000 mol wt. While the acidic chains are extensively modified, there must be little or no proteolysis of the basic δ-chains ($\delta_1$ and $\delta_2$).[189,192] Preliminary amino acid sequence data indicate that the amino-terminal region of the δ-chain of $F_{\alpha\beta}$ is identical to that of the δ-chain of intact F globulin.[117] After 4 days of germination $F_{\alpha\beta}$ also decreases, indicating further proteolysis, presumably also of the δ-chains.

Hara and Matsubara have partially characterized two enzymes that are involved in the initial stages of the proteolysis of the pumpkin globulin.[117,192] These enzymes were characterized on the basis of their activity toward the native globulin F, the partially degraded form $F_{\alpha\beta}$, and the individual γ, γ′, and δ (a mixture of $\delta_1$ and $\delta_2$) chains. The latter three substrates were obtained by the reduction and carboxymethylation of the native globulin F, followed by separation of the carboxymethyl (CM-) chains on DEAE-cellulose in the presence of urea.[66] Proteolysis of the substrates was monitored by SDS-PAGE, or by the release of ninhydrin-reactive, TCA-soluble peptides.

The first of these proteases (designated activity I) catalyzes the limited hydrolysis of the α and β subunits of globulin F to produce the 43,000 mol wt subunit representative of $F_{\alpha\beta}$. The activity is found in the ungerminated seed, and rapidly declines after 2 days of germination. Seeds imbibed in 5 m$M$ cycloheximide for 4 days were also found to contain activity I, at levels higher than those found in ungerminated seeds. In addition to hydrolyzing F to $F_{\alpha\beta}$ plus smaller peptide(s), the protease also hydrolyzes the CM-γ′, and the CM-γ-chain to a lesser extent. The enzyme is inactive towards $F_{\alpha\beta}$, the CM-δ chains, bovine serum albumin, ovalbumin, and horse heart cytochrome c. Activity I has a relatively narrow pH profile using F as the substrate, with an optimum of pH 6.8. It is partially inhibited by 5 m$M$ pCMB and 5 m$M$ EDTA, but is unaffected by PMSF, or pumpkin trypsin inhibitor.[117] When CM-γ′, rather than F, is used as the substrate, only EDTA is inhibitory. The enzyme has been found to be localized in the matrix regions of the protein bodies.[194] A similar F globulin specific protease has been described by Reilly et al.[195]

The second proteolytic activity, designated II, rapidly degrades $F_{\alpha\beta}$ to TCA-soluble peptides. CM-δ is also rapidly degraded, while F and CM-γ′ are more slowly hydrolyzed. Bovine serum albumin, ovalbumin, and cytochrome $c$ are hydrolyzed very slowly, or not at all. A pH optimum of approximately 6.5 was found for activity II with $F_{\alpha\beta}$ as the substrate. Activity II, absent from dry seeds and those germinated up to 2 days, rapidly increases to a maximal value by 4 days of germination, and then gradually declines thereafter. Protease II is sulfhydryl dependent, being activated by 2-mercaptoethanol and dithiothreitol, and strongly inhibited by pCMB; EDTA is also inhibitory, while $CoCl_2$ is slightly stimulatory; PMSF is without effect. It is unclear if protease II of Hara and Matsubara[117] is the same as the enzyme system II described by Reilly et al.,[195] though this seems quite likely.

The protein characterization data and enzymological data allow us to summarize the major stages in the degradation of the pumpkin storage globulin during germination:

$$F \xrightarrow{\text{Protease I}} F_{\alpha\beta} \xrightarrow{\text{Protease II}} \text{Peptides} \xrightarrow{\text{Protease III}} \text{Amino Acids}$$

(insoluble)   (soluble)

The first step, catalyzed by protease I, involved the limited proteolysis of globulin F, specifically at the acidic $\gamma$- and $\gamma'$-chains. In the second stage of mobilization, protease II extensively degrades the modified globulin $F_{\alpha\beta}$ to small peptides and possibly some free amino acids. The third stage, the degradation of the peptides produced in the second stage to free amino acids, has not been well characterized.

Hara and Matsubara[157] have described an arylamidase in pumpkin that hydrolyzes $N$-$\alpha$-benzoyl-D,L-arginine $p$-nitroanilide. The activity, initially absent in the dry seed, increases during germination, peaking at day 4, and gradually declining thereafter. The enzyme was purified 77-fold by ion exchange chromatography and gel filtration. An apparent molecular weight of 58,000 was found by gel filtration. The arylamidase appears to be sulfhydryl dependent, being activated approximately 20% by 1 m$M$ dithiothreitol or 2-mercaptoethanol. Further, it is inhibited by pCMB, not not by PMSF or EDTA. Both $F_{\alpha\beta}$ and CM-$\delta$ are hydrolyzed to TCA-soluble peptides and amino acids, while neither the native globulin (F) nor CM-$\gamma'$ are affected. When the reaction mixtures were examined by SDS-PAGE, extensive degradation of CM-$\gamma'$ to smaller peptides was observed. However, with $F_{\alpha\beta}$, much less degradation was noted, with little change in the apparent molecular weight of the modified globulin after digestion. This would indicate the removal of only a small part of the molecule as small peptides or amino acids. It seems likely that the pumpkin also contains a dipeptidase and one or more aminopeptidases active in the final stages of reserve protein mobilization. Such enzymes have been described in the closely related squash (*Cucurbita maxima*).[178,179]

## B. Proteolysis of Legume Storage Globulins

The proteolytic enzymes of many species of legumes have been examined at various levels of experimental sophistication. In spite of this, relatively little work has been published examining changes in the legume storage proteins by proteolysis during germination and early seedling growth. By far the best studied system in this regard is that of the germinating vetch (*Vicia sativa*) seed. In a series of papers Shutov and co-workers have examined the changes in the vetch legumin and vicilin during germination, as well as the proteolytic enzymes involved.

In examining the susceptibility of the vetch seed globulins to proteolysis by vetch seed extracts, Korolyova et al.[144] made a key observation. Neither legumin nor vicilin isolated from dry (ungerminated) seeds was attacked by extracts from ungerminated seeds. However, legumin and vicilin that had been purified from the cotyledons of seedlings $8^{1}/_{2}$ days old was readily converted by the extract of dry seeds into TCA-soluble products, as detected by reaction with TNBS. This suggests that during the course of germination and early seedling growth the storage globulins are in some manner modified, increasing their susceptibility to proteolysis.

The proteolytic enzymes involved in this initial modification of the storage proteins and their subsequent extensive degradation have been examined in considerable detail. The success in elucidating this system is due in large part to the use of a battery of protein and synthetic substrates, including the animal protein casein. It is, however, the use of both native and germination-modified forms of the vetch legumin and vicilin that has contributed to the solution of the puzzle. Eight different proteolytic enzymes have thus far been identified.

If extracts of ungerminated vetch seeds are chromatographed on DEAE-cellulose at pH

6.5, three peaks of activity capable of hydrolyzing casein to TCA-soluble products are eluted in the NaCl gradient.[122,132,150] None of the three peaks (D, C, and E, in order of elution) exhibit significant activity towards the storage proteins from ungerminated seeds.[122] However, both peaks C and D hydrolyze legumin from germinated seeds, while E hydrolyzes vicilin from germinated seeds. Similar chromatography of extracts of isolated protein bodies indicates that both C and D are found in the protein bodies, while E is apparently cytosolic.[132] If extracts of germinated cotyledons are examined, enzymes C and D are again observed, as are two additional peaks of activity, B and A, eluting from the column after C.[122,132] Both B and A hydrolyze casein. However, only A hydrolyzes the storage proteins from ungerminated vetch seeds.[122]

These proteases have each been characterized to varying degrees. Protease A has been purified 870-fold by a combination of chromatographic techniques.[118] While still heterogeneous by PAGE, the purified preparation appears to be free of other proteolytic activities. It is a sulfhydryl-dependent enzyme,[118,150] with an apparent molecular weight of approximately 16,000 to 25,000. It readily hydrolyzes native vetch vicilin and legumin, with a pH optimum of 4.6, yielding polypeptides with average lengths of 16 and 9 residues, respectively, upon prolonged incubation. No free amino acids appear to be produced, indicating A to be an endopeptidase.[118] Studies of the hydrolysis of the A- and B-chains of insulin suggest that protease A cleaves after the carboxyl group of aromatic and acidic amino acid residues. Protease A is absent during the first 4 days of germination, increasing thereafter to at least 8 days of growth.[132]

Protease B has been purified to homogeneity.[156] It exhibits a time course of development similar to that of protease A, and like A is sulfhydryl dependent.[132,150] It has an apparent molecular weight of approximately 25,000 to 38,000, depending upon the method used. Unlike protease A, it is inactive towards the native vetch legumin and vicilin, but readily hydrolyzes both after their initial partial proteolysis by A.[132] Further studies have indicated that hydrolysis by protease A of only one or two peptide bonds in the native legumin renders it maximally susceptible to the action of protease B.[197] Examination of the action of this protease on the two chains from insulin suggests that protease B specifically cleaves peptide bonds of the type Asn-X.[132] A pH optimum of 5.6 was found for the hydrolysis of the protease A-modified legumin by protease B.

Proteases C and D probably represent the vetch carboxypeptidases, although no definitive proof of this has yet been published.[132,150] They exhibit essentially constant activity throughout the first 8 days of germination and growth,[132] with the activity of protease C approximately four times higher than that of protease D.

Both proteases C and D hydrolyze the modified legumin, C, with a pH optimum of 5.7, D with an optimum of 6.8. Both are inhibited by DFP and only weakly or not at all by iodoacetate.[150] Both enzymes hydrolyze casein, but at a lower rate than the protease A-modified legumin.

Protease E remains poorly characterized. While present in the dormant seed, it rapidly declines upon germination, disappearing by $8^1/_2$ days of germination.[132,150] It hydrolyzes modified vicilin with a pH optimum of 5.5, but is inactive against the storage globulins of the ungerminated seed. It is a serine-dependent protease, being rapidly inactivated by DFP.

In addition to these proteases, three arylamidases with differing specificity have been characterized by Shutov et al.[15,174-176] Recently, the actual changes occurring in the vetch storage proteins during germination have been examined.[197] Using susceptibility to digestion by protease B as a probe, the initial modification of legumin appears to commence about day 2 or 3 of germination. By day 6 the modification appears to be complete. The legumin of the ungerminated vetch seed is typical in containing two classes of polypeptide chains. Two types of basic chains with molecular weights of approximately 26,000 and 23,000 may be demonstrated by ion exchange chromatography of the reduced legumin on DEAE- and

CM-cellulose, or by SDS-PAGE. Similarly, two types of acidic chains of 37,000 mol wt are found. Examination of the legumin from seeds germinated for 4 days indicates no change in the basic chains, while the acidic chains are extensively modified to polypeptides of 17,500 and 15,300 mol wt. Similar results were obtained with legumin treated with protease A in vitro.

The data present a picture very similar to that seen in the germinating pumpkin seed. The legumin of the dormant vetch seed does not serve as a substrate for the enzymes present in the dry seed, including the two carboxypeptidases (proteases C and D) and the three arylamidases. At least the first two of these enzymes are located in the same subcellular compartment as the storage protein, the protein bodies. By $4^{1}/_{2}$ days after imbibition endopeptidases A and B appear. Protease A initially inflicts a limited number of proteolytic cleavages upon the legumin acidic chains. While not significantly changing the overall physical properties of the legumin, this initial proteolysis renders it susceptible to further attack by both proteases A and B. By 5 days after the beginning of imbibition, extensive hydrolysis of the legumin acidic chains (with the loss of approximately 19 peptides) has occurred, reducing them to approximately one half their original size. Continued action of the endopeptidases further degrades the modified legumin, including the more resistant basic chains, to small peptides. These in turn are degraded to free amino acids by the carboxypeptidases in the protein bodies, and possibly also by cytosolic peptidases.

The nature of the degradation of vicilin during germination is less clear. However, it seems likely that its degradation follows a pathway similar to that of legumin. It is at present unclear whether the initial proteolysis of vicilin gives rise to intermediate forms with significant life times, such as those seen during legumin degradation. It may be noted that such intermediate forms have been noted in other species of legumes.

The mode of legumin and vicilin degradation in the vetch would appear to be typical of the legumes in general. The initial changes in the vetch legumin are very similar to those observed with soybean legumin, glycinin.[198] We have observed the rapid degradation of the acidic chains, with the $A_3$-chains being degraded much more rapidly than the other acidic chains ($A_1$, $A_2$, and $A_4$).[61] As in the vetch, the acidic chains appear to be hydrolyzed from their initial molecular weights of 37,000 to 45,000 to forms of approximately 21,000 to 30,000. While the initial acidic chain forms have largely disappeared by day 6 after the start of imbibition, the basic chains persist until day 8 before suffering extensive degradation. The different subunits of the soybean vicilin — conglycinin — are also degraded at different rates, with the α- and α'-subunits being degraded more rapidly than the β subunit. This same pattern of degradation of glycinin and β-conglycinin has been observed in vitro when extracts of ungerminated soybeans are incubated at pH 8.0. This degradation has been ascribed to a metalloproteinase present in the ungerminated seed.[136]

The appearance during germination of sulfhydryl-dependent endopeptidase activity has been noted in a number of legume species, including the mung bean (*Vigna radiata*),[72] pea (*Pisum sativum*),[9] common bean (*Phaseolus vulgaris*),[154] and the peanut (*Arachis hypogaea*).[123] The enzyme from the mung bean, vicilin peptidohydrolase, has been extensively studied by Chrispeels and co-workers.[72,120,145,158,159,199,200] It is active against the mung bean vicilin, which consitutes the major storage globulin in this legume species,[201] with a pH optimum of 5.1. Its initial action on the vicilin is a limited specific proteolysis, converting the vicilin (made up predominantly of subunits, $M_r$ 50,000) to 20,000 and 30,000 mol wt products.[72] With prolonged incubation, these too are degraded to smaller peptides. The action of the enzyme on the mung bean legumin has not been described. A similar pattern of vicilin degradation has been noted in vivo in the bean *P. vulgaris*.[80] The vicilin peptidohydrolase has been shown to appear during germination due to *de novo* synthesis,[159] and has been demonstrated to be contained within the protein bodies of germinated seeds.[158]

## C. Proteolysis of Legume Storage Proteins Other Than Globulins — Lectins and Trypsin Inhibitors

While most studies of protein mobilization during germination have concentrated upon the globulins of the legumin and vicilin type, the degradation of several other storage proteins has been examined. Murray[80] examined the mobilization of glycoproteins in the germinating kidney bean (*P. vulgaris*) by a combination of electrophoretic and immunological techniques. No significant decrease in total cotyledonary protein was noted until day 6 after the start of imbibition. This decrease was primarily associated with a decrease in the pH 4.7 insoluble fraction, i.e., the kidney bean vicilin "euphaseolin". This decline in the vicilin was accompanied by the conversion of the three subunits of the vicilin ($M_r$ 54,000, 49,000, and 46,000) into degradation products of 23,000 and 21,000 mol wt. While large amounts of the euphaseolin persisted even to 19 days after imbibition, the kidney bean lectin, phytohemagglutinin, declined much more rapidly. Most of the lectin had disappeared by day 6, and was undetectable by day 12. SDS-PAGE suggests that the 30,000 mol wt protomeric subunit of the lectin is degraded to products with molecular weights of 18,500, 16,500, 15,000, and 13,000, though this conclusion should be viewed with some caution.

Murray has suggested that the rapid hydrolysis of the lectin might be expected if its localization is in the cytoplasm rather than the protein bodies. The rapid degradation of cytosolic proteins during the early stages of germination has been observed in other systems also. Manickam and Carlier[109] have described a 12,000 mol wt protein present in the embryonic axis of the mung bean seed. This is rapidly degraded during germination, disappearing by 24 hr after the beginning of imbibition. This protein has been demonstrated to be cytosolic in localization. A similar rapidly degraded low-molecular-weight polypeptide has also been described in the radicle of pea seeds.[110]

The Bowman-Birk type trypsin inhibitor of the mung bean is another cytosolic protein that exhibits a more rapid degradation than the protein body-bound storage proteins. When mung bean seeds are germinated there is a 3-day lag period after the start of imbibition before there is significant degradation of the bulk portion of the storage protein. Over this same period, the amount of trypsin inhibitor activity in the seed remains essentially constant.[134] However, during this first 72 hr the trypsin inhibitor undergoes a series of proteolytic modifications, which, while preserving the inhibitor activity, results in the degradation of nearly 20% of the inhibitor molecule.[104] Lorensen et al.[104] have examined this proteolysis using chromatography on Sephadex® G-75 and DEAE-Sephadex®, and PAGE. The data suggest a specific pathway for the initial stages of the degradation of the mung bean trypsin inhibitor during germination:

$$F \rightarrow E \rightarrow C \rightarrow A$$

where F is the major form of the inhibitor found in the quiescent dry seed, and E, C, and A are forms that appear during germination. Under the growth conditions used, inhibitor F had nearly disappeared from the seed by 48 hr after the start of imbibition. A number of these inhibitor forms were subsequently characterized by Wilson and Chen.[182] The mung bean trypsin inhibitor F is a typical Bowman-Birk type trypsin inhibitor with 80 amino acid residues in a single polypeptide chain. Inhibitor E, the first observable degradation product, differs from F by the loss of the carboxyl-terminal tetrapeptide sequence -Lys-Asp-Asp-Asp. Inhibitor C is obtained from E by the loss of an additional two carboxyl-terminal residues, -Met-Asp, the loss of the amino-terminal octapeptide sequence Ser-Ser-His-His-His-Asp-Ser-Ser-, and an internal cleavage at the Ala$^{35}$-Asp$^{36}$ peptide bond.

We have now examined some of the basic properties of the enzymatic system catalyzing the initial degradation of the inhibitor.[134] An assay for inhibitor modifying proteolytic activity was developed based upon the electrophoretic separation of inhibitors F, E, and C. Using crude extracts from ungerminated and germinated seeds, a pH optimum of pH 4 was found for both the conversion of inhibitor F to E and E to C. Extracts of ungerminated seeds catalyze the former, but not the latter, reaction. The activity catalyzing the conversion E to C appears on day 2 after imbibition, and increases up to at least day 6. A number of organic reagents were tested for their ability to inhibit the inhibitor degrading enzymes. Iodoacetate, pepstatin A, leupeptin, and EDTA were without effect on the F to E and E to C reactions. PMSF was found to strongly inhibit the E to C reaction, although it had little or no effect on the F to E reaction.

These data suggest that the major endopeptidase of the germinating mung bean, vicilin peptidohydrolase, is not necessary in these early stages of inhibitor degradation, as this enzyme has been shown to be inhibited by both iodoacetate and leupeptin.[72,202] This conclusion was confirmed by gel filtration of an extract of cotyledons from beans germinated for 5 days. Three peaks of activity hydrolyzing azocasein at pH 5.7 were found, with apparent molecular weights of greater than 130,000, approximately 60,000, and 26,000. The last peak was shown to be the vicilin peptidohydrolase on the basis of its molecular weight and sensitivity to iodoacetate and leupeptin. All of the activity catalyzing the E to C reaction was found to be associated with the second proteolytic peak, although the reaction F to E was catalyzed by all three activity peaks.[134]

The limited specific proteolysis of a Bowman-Birk type trypsin inhibitor is not unique to the mung bean. A similar degradation at the amino-terminus has also been demonstrated in the adzuki bean (*Vigna angularis*)[203] and is also suggested in the data from the soybean.[105] Soybeans also contain a second type of trypsin inhibitor, the Kunitz soybean trypsin inhibitor (see Section III.D). During germination and seedling growth, the Kunitz inhibitor also undergoes a specific proteolytic modification that alters its electrophoretic mobility[92,105,204,205] and chromatographic properties.[105] This modification occurs at a much slower rate than that of the Bowman-Birk inhibitors in the same seed.[105] We have examined the native and modified forms of the Kunitz inhibitor from the Fiskeby V cultivar of soybeans. This cultivar contains the $Ti^b$ form of the inhibitor. The modification appears to result from the loss of approximately ten amino acid residues from the carboxyl-terminus of the native protein.[204] The Kunitz inhibitor does not appear to be significantly degraded past this point, even after 12 days after imbibition.[105,205] It thus seems unlikely that the Kunitz inhibitor functions primarily as a storage protein in the germinating seed.

## VI. THE GENERALIZED PATHWAY FOR STORAGE PROTEIN MOBILIZATION DURING GERMINATION

The data examined above suggest a general pathway for the hydrolysis of storage proteins in the dicots (Figure 1) essentially the same as that already proposed for the monocots.[5] The quiescent dry seed contains, in addition to the storage proteins, a number of proteolytic enzymes, including several arylamidases and alkaline peptidases, possibly one or more endopeptidases in relatively low quantities, and often one or more acid carboxypeptidases. Proteolysis of the storage protein does not occur in the mature, dessicating seed, the quiescent dry seed, or the newly imbibed seed for one or more of the following reasons:

1. The native storage protein does not serve as a substrate for the proteolytic enzymes present.
2. The enzyme(s) are localized in the cytoplasm in a compartment other than the one in which the storage protein is contained.
3. Dessication has reversibly inactivated the enzyme(s).

The degradation of the storage protein is initiated by an endopeptidase which may exhibit

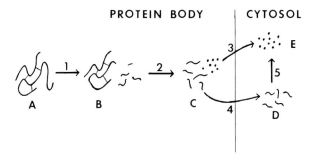

FIGURE 1. Generalized pathway for the mobilization of storage proteins contained in protein bodies. The substrates and products on the pathway are (A) native storage protein; (B) storage protein modified by limited specific proteolysis, plus resulting oligo- and polypeptide fragments; (C) oligopeptides and free amino acids; (D) oligopeptides now in the cytosol; (E) free amino acids in the cytosol. A, B, and C are localized within the protein body. The reactions and enzymes involved are (1) limited specific proteolysis (typically by a sulfhydryl-dependent endopeptidase); (2) further degradation by enzyme 1 and/or another endopeptidase, and possibly by carboxypeptidase(s) in the protein body; (3 and 4) transport amino acids and oligopeptides, respectively, across the membrane of the protein body into the cytosol; (5) final degradation of oligopeptides by cytosolic peptidases.

a relatively high specificity for the native storage protein. This endopeptidase inflicts a limited number of specific proteolytic cleavages on the storage protein, producing the modified storage protein. This initial modification renders the storage protein susceptible to further endoproteolytic attack, either by the initiating enzyme, or by another endopeptidase, or both. In addition, the modified storage protein may now be susceptible to the action of any carboxypeptidase present in the protein body. The net effect of these first two stages of mobilization is the degradation of the storage protein contained in the protein body into oligopeptides and free amino acids. Further degradation of the oligopeptides to amino acids continues through the action of the carboxypeptidases within the protein body, and/or the arylamidases and alkaline peptidases in the cytosol.

The degradation of cytosolic storage proteins such as the Bowman-Birk proteinase inhibitors apparently follows a similar pathway. Again, a limited specific proteolysis is necessary to render the protein susceptible to further degradation.

## ACKNOWLEDGMENT

Most of the work noted here from the author's laboratory was supported by grants PCM-8003854 and PCM-8301202 from the National Science Foundation.

## REFERENCES

1. **Ashton, F. M.,** Mobilization of storage proteins of seeds, *Annu. Rev. Plant Physiol.*, 27, 95, 1976.
2. **Davies, H. V. and Slack, P. T.,** The control of food mobilization in seeds of dicotyledonous plants, *New Phytol.*, 88, 41, 1981.
3. **Koller, D., Mayer, A. M., Poljakoff-Mayber, A., and Klein, S.,** Seed germination, *Annu. Rev. Plant Physiol.*, 13, 437, 1962.
4. **Ryan, C. A.,** Proteolytic enzymes and their inhibitors in plants, *Annu. Rev. Plant Physiol.*, 24, 173, 1973.

5. **Mikola, J.**, Proteinases, peptidases, and inhibitors of endogenous peptidases in germinating seeds, in *Seed Proteins*, Daussant, J., Mosse, J., and Vaughan, J., Eds., Academic Press, New York, 1983, chap. 2.
6. **Mayer, A. M. and Marbach, I.**, Biochemistry of the transition from resting to germinating state in seeds, *Prog. Phytochem.*, 7, 95, 1981.
7. **Bewley, J. D. and Black, M.**, *Physiology and Biochemistry of Seeds in Relation to Germination*, Springer-Verlag, New York, 1978, chap. 6.
8. **Daussant, J., Neucere, N. J., and Conkerton, E. J.**, Immunochemical studies on *Arachis hypogaea* proteins with particular reference to the reserve proteins. II. Protein modification during germination, *Plant Physiol.*, 44, 480, 1969.
9. **Basha, S. M. M. and Beevers, L.**, The development of the proteolytic activity and protein degradation during germination of *Pisum sativum* L., *Planta*, 124, 77, 1975.
10. **Juo, P.-S. and Stotzky, G.**, Changes in protein spectra of bean seed during germination, *Can. J. Bot.*, 48, 1347, 1970.
11. **Pusztai, A., Croy, R. R. D., Grant, G., and Watt, W. B.**, Compartmentalization in the cotyledonary cells of *Phaseolus vulgaris* L. seeds: a differential sedimentation study, *New Phytol.*, 79, 61, 1977.
12. **Catsimpoolas, N., Campbell, T. G., and Meyer, E. G.**, Immunochemical study of changes in reserve proteins of germinating soybean seeds, *Plant Physiol.*, 43, 799, 1968.
13. **Catsimpoolas, N., Ekenstam, C., Rogers, D. A., and Meyer, E. G.**, Protein subunits in dormant and germinating soybean seeds, *Biochim. Biophys. Acta*, 168, 122, 1968.
14. **Kumar, K. G. and Venkataraman, L. V.**, Chickpea seed proteins: modification during germination, *Phytochemistry*, 17, 605, 1978.
15. **Shutov, A. D. and Vaintraub, I. A.**, Primary changes of reserve proteins during germination of vetch seeds, *Fiziol. Rast.*, 20, 504, 1973.
16. **Kumar, K. G., Venkataraman, L. V., and Rao, A. G. A.**, Chickpea seed proteins: conformational changes in the 10.3S protein during germination, *J. Agric. Food Chem.*, 28, 518, 1980.
17. **Cooper, T. G., Ed.**, *Tools of Biochemistry*, John Wiley & Sons, New York, 1977, chap. 6.
18. **Weber, K. and Osborne, M.**, The reliability of molecular weight determinations by dodecyl sulfate-polyacrylamide gel electrophoresis, *J. Biol. Chem.*, 244, 4406, 1969.
19. **Laemmli, U. K.**, Cleavage of structural proteins during the assembly of the head of bacteriophage T4, *Nature (London)*, 227, 680, 1970.
20. **Casey, R.**, Genetic variability in the structure of the α-subunits of legumin from *Pisum* — a two dimensional gel electrophoresis study, *Heredity*, 43, 265, 1979.
21. **Staswick, P. E., Hermodson, M. A., and Nielsen, N. C.**, Identification of the acidic and basic subunit complexes of glycinin, *J. Biol. Chem.*, 256, 8752, 1981.
22. **Kitamura, K., Takaji, T., and Shibasaki, K.**, Subunit structure of soybean 11S globulin, *Agric. Biol. Chem.*, 40, 1837, 1976.
23. **Barton, K. A., Thompson, J. F., Madison, J. T., Rosenthal, R., Jarvis, N. P., and Beachy, R. N.**, The biosynthesis and processing of high molecular weight precursors of soybean glycinin subunits, *J. Biol. Chem.*, 257, 6089, 1982.
24. **Croy, R. R. D., Gatehouse, J. A., Evans, I. M., and Boulter, D.**, Characterisation of the storage protein subunits synthesized *in vitro* by polyribosomes and RNA from developing pea (*Pisum sativum* L.). I. Legumin, *Planta*, 148, 49, 1980.
25. **O'Farrell, P. H.**, High resolution two-dimensional electrophoresis of proteins, *J. Biol. Chem.*, 250, 4007, 1975.
26. **Casey, R.**, Immunoaffinity chromatography as a means of purifying legumin from *Pisum* (pea) seeds, *Biochem. J.*, 177, 509, 1979.
27. **Gatehouse, J. A., Croy, R. R. D., and Boulter, D.**, Isoelectric focusing properties and carbohydrate content of pea (*Pisum sativum*) legumin, *Biochem. J.*, 185, 497, 1980.
28. **Krishna, T. G., Croy, R. R. D., and Boulter, D.**, Heterogeneity in subunit composition of the legumin of *Pisum sativum*, *Phytochemistry*, 18, 1879, 1979.
29. **Lei, M.-G., Tyrell, D., Bassette, R., and Reeck, G. R.**, Two-dimensional electrophoretic analysis of soybean proteins, *J. Agric. Biol. Chem.*, 31, 963, 1983.
30. **Matta, N. K., Gatehouse, J. A., and Boulter, D.**, Molecular and subunit heterogeneity of legumin of *Pisum sativum* L. (garden pea) — a multidimensional gel electrophoretic study, *J. Exp. Bot.*, 32, 1295, 1981.
31. **Hu, B. and Esen, A.**, Heterogeneity of soybean proteins. Two dimensional electrophoretic maps of three solubility fractions, *J. Agri. Biol. Chem.*, 30, 21, 1982.
32. **Burnette, W. N.**, "Western blotting": electrophoretic transfer of proteins from sodium dodecyl sulfate — polyacrylamide gels to unmodified nitrocellulose and radiographic detection with antibody and radioiodinated protein A, *Anal. Biochem.*, 112, 195, 1981.

33. **Symington, J., Green, M., and Broekmann, K.,** Immunoautographic detection of proteins after electrophoretic transfer from gels to diazo-paper: analysis of adenovirus encoded proteins, *Proc. Natl. Acad. Sci. U.S.A.*, 78, 177, 1981.
34. **Renart, J., Reiser, J., and Stark, G. R.,** Transfer of protein from gels to diazobenzyloxymethyl — paper and detection with antisera: a method for studying antibody specificity and antigen structure, *Proc. Natl. Acad. Sci. U.S.A.*, 76, 3116, 1979.
35. **Osborne, T. B., Ed.,** *The Vegetable Proteins*, 2nd ed., Longmans, Green, London, 1924.
36. **Bewley, J. D. and Black, M., Eds.,** *Physiology and Biochemistry of Seeds in Relation to Germination*, Springer-Verlag, New York, 1978, chap. 2.
37. **Danielson, C. E.,** Plant proteins, *Annu. Rev. Plant Physiol.*, 7, 215, 1956.
38. **Danielson, C. E.,** Seed globulins of the Gramineae and Leguminosae, *Biochem. J.*, 44, 387, 1949.
39. **Millerd, A.,** Biochemistry of legume seed proteins, *Annu. Rev. Plant Physiol.*, 26, 53, 1975.
40. **Pernollet, J.-C.,** Protein bodies of seeds: ultrastructure, biochemistry, biosynthesis and degradation, *Phytochemistry*, 17, 1473, 1978.
41. **Altshul, A. M., Yatsu, L. Y., Ory, R. L., and Engleman, E. M.,** Seed proteins, *Annu. Rev. Plant Physiol.*, 17, 113, 1966.
42. **Klein, S. and Ben-Shaul, Y.,** Changes in cell fine structure of lima bean axes during early germination, *Can. J. Bot.*, 44, 331, 1966.
43. **Yoo, B. Y.,** Ultrastructural changes in cells of pea embryo radicles during germination, *J. Cell Biol.*, 45, 158, 1970.
44. **Perner, E.,** Elektronmikroskopische untersuchungen an zellen von embryonen im zustand volliger samenruhe, *Planta*, 67, 324, 1965.
45. **Youle, R. J. and Huang, A. H. C.,** Protein bodies from the endosperm of castor bean. Subfractionation, protein components, lectins, and changes during germination, *Plant Physiol.*, 58, 703, 1976.
46. **Tully, R. E. and Beevers, H.,** Protein bodies of castor bean endosperm. Isolation, fractionation, and the characterization of protein components, *Plant Physiol.*, 58, 710, 1976.
47. **Briarty, L. G., Coult, D. A., and Boulter, D.,** Protein bodies of germinating seeds of *Vicia faba*, *J. Exp. Bot.*, 21, 513, 1970.
48. **Öpik, H.,** Changes in cell fine structure in the cotyledons of *Phaseolus vulgaris* L. during germination, *J. Exp. Bot.*, 17, 427, 1966.
49. **Tombs, M. P.,** Protein bodies of the soybean, *Plant Physiol.*, 42, 797, 1967.
50. **Swift, J. M. and Buttrose, M. S.,** Protein bodies, lipid layers, and amyloplasts in freeze-etched pea cotyledons, *Planta*, 109, 61, 1973.
51. **Bagley, B. W., Cherry, J. H., Rollins, M. L., and Altschul, A. M.,** A study of protein bodies during germination of peanut (*Arachis hypogaea*) seed, *Am. J. Bot.*, 50, 523, 1963.
52. **St. Angelo, A. J., Yatsu, L. Y., and Altschul, A. M.,** Isolation of edestin from aleurone grains of *Cannabis sativa*, *Arch. Biochem. Biophys.*, 124, 199, 1968.
53. **Dure, L. S. and Waters, L.,** Long-lived messenger RNA: evidence from cotton seed germination, *Science*, 147, 410, 1965.
54. **Derbyshire, E., Wright, D. J., and Boulter, D.,** Legumin and vicilin, storage proteins of legume seeds, *Phytochemistry*, 15, 3, 1976.
55. **Dieckert, J. W. and Dieckert, M. C.,** The comparative anatomy of the principal reserve proteins of seeds, in *Seed Proteins of Dicotyledonous Plants*, Muntz, K., Ed., Akademie-Verlag, Berlin, 1979, 73.
56. **Boulter, D., Derbyshire, E., and Croy, R.,** Structure and function of the storage proteins of some economically important legume seeds, in *Seeds Proteins of Dicotyledonous Plants*, Müntz, K., Ed., Akademie-Verlag, Berlin, 1979, 11.
57. **Bollini, R. and Chrispeels, M. J.,** Characterization and subcellular localization of vicilin and phytohemagglutinin, the two major reserve proteins of *Phaseolus vulgaris* L., *Planta*, 142, 291, 1978.
58. **Catsimpoolas, N. and Ekenstam, C.,** Isolation of alpha, beta, and gamma conglycinins, *Arch. Biochem. Biophys.*, 129, 490, 1969.
59. **Bradley, R. A., Atkinson, D., Hauser, H., Oldani, D., Green, J. P., and Stubbs, J. M.,** The structure, physical and chemical properties of the soy bean protein glycinin, *Biochim. Biophys. Acta*, 412, 214, 1975.
60. **Kitamura, K. and Shibasaki, K.,** Isolation and some physicochemical properties of the acidic subunits of soybean 7S globulin, *Agric. Biol. Chem.*, 39, 945, 1975.
61. **Moreira, M. A., Hermodson, M. A., Larkins, B. A., and Nielsen, N. C.,** Partial characterization of the acidic and basic polypeptides of glycinin, *J. Biol. Chem.*, 254, 9921, 1979.
62. **Ereken-Tumer, N., Thanh, V. H., and Nielsen, N. C.,** Purification and characterization of mRNA from soybean seeds. Identification of glycinin and β-conglycinin precursors, *J. Biol. Chem.*, 256, 8756, 1981.
63. **Ereken-Tumer, N., Richter, J. D., and Neilsen, N. C.,** Structural characterization of glycinin precursors, *J. Biol. Chem.*, 257, 4016, 1981.

64. **Barton, K. A., Thompson, J. F., Madison, J. T., Rosenthal, R., Jarvis, N. P., and Beachy, R. N.,** The biosynthesis and processing of high molecular weight precursors of soybean glycinin subunits, *J. Biol. Chem.*, 257, 6089, 1982.
65. **Chrispeels, M. J., Higgins, T. J. V., and Spencer, D.,** Assembly of storage protein oligomers in the endoplasmic reticulum and processing of the polypeptides in the protein bodies of developing pea cotyledons, *J. Cell Biol.*, 93, 306, 1982.
65a. **Dlouha, V., Kiel, B., and Sorm, F.,** Structure of peptides isolated from the tryptic hydrolysate of the A chain of edestin, *Coll. Czech. Chem. Commun.*, 29, 1835, 1963.
66. **Hara, I., Wada, K., Wakabayashi, S., and Matsubara, H.,** Pumpkin (*Cucurbita* sp.) seed globulins. I. Purification, characterization, and subunit structure, *Plant Cell Physiol.*, 17, 799, 1976.
67. **Koshiyama, I. and Fukushima, D.,** Identification of the 7S globulin with β-conglycinin, *Phytochemistry*, 15, 157, 1976.
68. **Thanh, V. H. and Shibasaki, K.,** Major proteins of soybean seeds. Subunit structure of β-conglycinin, *J. Agric. Food Chem.*, 26, 692, 1978.
69. **Thanh, V. H. and Shibasaki, K.,** Heterogeneity of β-conglycinin, *Biochim. Biophys. Acta*, 439, 326, 1976.
70. **Thanh, V. H. and Shibasaki, K.,** Beta-conglycinin from soybean proteins. Isolation and immunological and physicochemical properties of the monomeric forms, *Biochim. Biophys. Acta*, 490, 370, 1977.
71. **Sykes, G. E., and Gayler, K. R.,** Detection and characterization of a new β-conglycinin, *Arch. Biochem. Biophys.*, 210, 525, 1981.
72. **Baumgartner, B. and Chrispeels, M. J.,** Purification and characterization of vicilin peptidohydrolase, the major endopeptidase in the cotyledons of mung bean seedlings, *Eur. J. Biochem.*, 77, 223, 1977.
73. **Lis, H., Sharon, N., and Katchalski, E.,** Soybean hemagglutinin, a plant glycoprotein. I. Isolation of a glycopeptide, *J. Biol. Chem.*, 241, 684, 1966.
74. **Kocourek, J. and Hořejší, V.,** Defining a lectin, *Nature (London)*, 290, 188, 1981.
75. **Pusztai, A., Croy, R. R. D., Grant, G., and Steward, J, C.,** Seed lectins: distribution, location, and biological role, in *Seed Proteins*, Daussant, J., Mosse, J., and Vaughan, J., Eds., Academic Press, New York, 1983, chap. 3.
76. **Goldstein, I. J. and Hayes, C. E.,** Lectins: carbohydrate-binding proteins, *Adv. Carbohydr. Chem. Biochem.*, 35, 127, 1978.
77. **Sumner, J. B. and Howell, S. F.,** The identification of the hemagglutinin of the jack bean with concanavalin A, *J. Bacteriol.*, 32, 227, 1936.
78. **Liener, I. E. and Pallansch, M. J.,** Purification of a toxic substance from defatted soy bean flour, *J. Biol. Chem.*, 197, 29, 1952.
79. **Pusztai, A., Clarke, E. M. W., King, T. B., and Stewart, J. C.,** Nutritional evaluation of kidney bean (*Phaseolus vulgaris*): chemical composition, lectin content and nutritional value of selected cultivars, *J. Sci. Food Agric.*, 30, 843, 1979.
80. **Murray, D. R.,** Proteolysis in cotyledon cells of *Phaseolus vulgaris*. II. Mobilization of glycoproteins following germination, *Z. Pflanzenphysiol.*, 108, 17, 1982.
81. **Barker, R. D. J., Derbyshire, E., Yarwood, A., and Boulter, D.,** Purification and characterization of the major storage proteins of *Phaseolus vulgaris* seeds, and their intracellular and cotyledonary distribution, *Phytochemistry*, 15, 751, 1976.
82. **Horisberger, M. and Vonlanthen, M.,** Ultrastructural localization of soybean agglutinin on thin sections of *Glycine max* soybean var. Altona by the gold method, *Histochemistry*, 65, 181, 1980.
83. **Clarke, A. E., Knox, R. B., and Jermyn, M. A.,** Localization of lectins in legume cotyledons, *J. Cell Biol.*, 19, 157, 1975.
84. **Laskowski, M., Jr. and Kato, I.,** Protein inhibitors of proteinases, *Annu. Rev. Biochem.*, 49, 593, 1980.
85. **Richardson, M.,** The proteinase inhibitors of plants and microorganisms, *Phytochemistry*, 16, 159, 1977.
86. **Wilson, K. A.,** The structure, function, and evolution of legume proteinase inhibitors, in *Antinutrients and Natural Toxicants in Food*, Ory, R. L., Ed., Food and Nutrition Press, Westport, Conn., 1981, chap. 11.
87. **Weder, J. K. P.,** Protease inhibitors in the Leguminosae, in *Advances in Legume Systematics*, Polhill, R. M. and Raven, P. H., Eds., Royal Botanical Garden, Kew, England, 1981, 533.
88. **Millar, D. B. S., Willick, G. E., Steiner, R. F., and Fratalli, V.,** Soybean inhibitors. IV. The reversible self-association of a soybean proteinase inhibitor, *J. Biol. Chem.*, 244, 281, 1969.
89. **Kunitz, M.,** Crystalline soybean trypsin inhibitor, *J. Gen. Physiol.*, 29, 149, 1946.
90. **Koide, T. and Ikenaka, T.,** Studies on soybean trypsin inhibitors. III. Amino-acid sequence of the carboxyl-terminal region and the complete amino-acid sequence of soybean trypsin inhibitor (Kunitz), *Eur. J. Biochem.*, 32, 417, 1973.
91. **Singh, L., Wilson, C. M., and Hadley, H. H.,** Genetic differences in soybean trypsin inhibitors separated by disc gel electrophoresis, *Crop Sci.*, 9, 489, 1969.
92. **Orf, J. H. and Hymowitz, T.,** Genetics of the Kunitz trypsin inhibitor: an antinutritional factor in soybeans, *J. Am. Oil Chem. Soc.*, 56, 722, 1979.

93. **Kortt, A. A. and Jermyn, M. A.,** *Acacia* proteinase inhibitors. Purification and properties of the trypsin inhibitors from *Acacia elata* seed, *Eur. J. Biochem.,* 115, 551, 1981.
94. **Yamamoto, M., Hara, S., and Ikenaka, T.,** Amino acid sequences of two trypsin inhibitors from wing bean seeds (*Psophocarpus tetragonolobus* [L.] DC.), *J. Biochem.,* 94, 849, 1983.
95. **Joubert, F. J.,** Purification and some properties of the proteinase inhibitors from *Acacia sieberana* (paperbark acacia) seed, *Phytochemistry,* 22, 53, 1983.
96. **Odani, S., Ono, T., and Ikenaka, T.,** The reactive site amino acid sequences of silk tree (*Albizzia julibrissin*) seed proteinase inhibitors, *J. Biochem.,* 88, 297, 1980.
97. **Chrispeels, M. J. and Baumgartner, B.,** Trypsin inhibitor in mung bean cotyledons. Purification, characteristics, subcellular localization and metabolism, *Plant Physiol.,* 61, 617, 1978.
98. **Miege, M. N., Mascherpa, J. M., Royer-Spierer, A., Grange, A., and Miege, J.,** Analyse des corps proteique isoles de *Lablab purpureus* (L.) Sweet: localization intracellulaire des globulines, proteases, et inhibiteurs de la trypsine, *Planta,* 131, 81, 1976.
99. **Hobday, S. M., Thurman, D. A., and Barker, D. J.,** Proteolytic and trypsin inhibitory activities in extracts of germinating *Pisum sativum* seeds, *Phytochemistry,* 12, 1041, 1973.
100. **Ryan, C. A. and Green, T. R.,** Proteinase inhibitors in natural plant protection, *Recent Adv. Phytochem.,* 8, 123, 1974.
101. **Pusztai, A.,** Metabolism of trypsin-inhibitory proteins in the germinating seeds of kidney beans (*Phaseolus vulgaris*), *Planta,* 107, 121, 1972.
102. **Filho, J. X.,** Trypsin inhibitors during germination of *Vigna sinensis* seeds, *Physiol. Plant.,* 28, 149, 1973.
103. **Baumgartner, B. and Chrispeels, M. J.,** Partial characterization of a protease inhibitor which inhibits the major endopeptidase present in the cotyledons of mung beans, *Plant Physiol.,* 58, 1, 1976.
104. **Lorensen, E., Prevosto, R., and Wilson, K. A.,** The appearance of new active forms of trypsin inhibitor in germinating mung bean (*Vigna radiata*) seeds, *Plant Physiol.,* 68, 88, 1981.
105. **Tan-Wilson, A. L., Rightmire, B. R., and Wilson, K. A.,** Different rates of metabolism of soybean proteinase inhibitors during germination, *Plant Physiol.,* 70, 493, 1982.
106. **Booth, A. N., Robbins, D. J., Ribelin, W. E., and DeEds, F.,** Effects of raw soybean meal and amino acids on pancreatic hypertrophy in rats, *Proc. Soc. Exp. Med.,* 104, 681, 1960.
107. **Liener, I. E.,** The nutritional significance of plant protease inhibitors, *Proc. Nutr. Soc.,* 38, 109, 1979.
108. **Youle, R. J. and Huang, A. H. C.,** Albumin storage proteins in the protein bodies of castor bean, *Plant Physiol.,* 61, 13, 1978.
108a. **Sharief, F. S. and Li, S. S.-L.,** Amino acid sequence of small and large subunits of seed storage protein from *Ricinus communis*, *J. Biol. Chem.,* 257, 14753, 1982.
108b. **Odani, S., Koide, T., and Ohnishi, K.,** Structural relationship between barley (*Hordeum vulgare*) trypsin inhibitor and castor bean (*Ricinus communis*) storage protein, *Biochem. J.,* 213, 543, 1983.
109. **Manickam, A. and Carlier, A. R.,** Isolation and function of a low molecular weight protein of mung bean embryonic axes, *Planta,* 149, 234, 1980.
110. **Murray, D. R.,** Proteolysis in the axis of the germinating pea seed. II. Changes in polypeptide composition, *Planta,* 147, 117, 1979.
110a. **Hartley, B. S.,** Proteolytic enzymes, *Annu. Rev. Biochem.,* 29, 45, 1960.
111. **Mikola, J.,** Functions of different plant peptidases in germinating seeds, in *Seed Proteins of Dicotyledonous Plants,* Müntz, K., Ed., Akademie-Verlag, Berlin, 1979, 125.
112. **Anson, M. L.,** The estimation of pepsin, trypsin, papain, and cathepsin with hemoglobin, *J. Gen. Physiol.,* 22, 79, 1938.
113. **Kunitz, M.,** Crystalline soybean trypsin inhibitor. II. General properties, *J. Gen. Physiol.,* 30, 291, 1947.
114. **Satake, K., Okuyama, T., Ohashi, M., and Shinoda, T.,** The spectrophotometric determination of amine, amino acid, and peptide with 2,4,6-trinitrobenzene 1-sulfonic acid, *J. Biochem.,* 47, 654, 1960.
115. **Charney, J. and Tomarelli, R. N.,** A colorimetric method for the determination of proteolytic activity of duodenal juice, *J. Biol. Chem.,* 171, 501, 1947.
116. **Lin, Y., Means, G. E., and Feeney, R. E.,** The action of proteolytic enzymes on N,N-dimethyl proteins. Basis for a microassay for proteolytic enzymes, *J. Biol. Chem.,* 244, 789, 1969.
117. **Hara, I. and Matsubara, H.,** Pumpkin (*Cucurbita* sp.) seed globulin. V. Proteolytic activities involved in globulin degradation in ungerminated seeds, *Plant Cell Physiol.,* 21, 219, 1980.
118. **Bulmaga, V. P. and Shutov, A. D.,** Partial purification and characterization of protease A of germinating vetch seeds, hydrolyzing native reserve proteins, *Biokhimiya,* 42, 1983, 1977.
119. **Minamikawa, T.,** Hydrolytic enzyme activities and degradation of storage components in cotyledons of germinating *Phaseolus mungo* seeds, *Bot. Mag. Tokyo,* 92, 1, 1979.
120. **Chrispeels, M. J. and Boulter, D.,** Control of storage protein metabolism in the cotyledons of germinating mung beans: role of endopeptidase, *Plant Physiol.,* 55, 1031, 1975.
121. **Tomomatsu, A., Iwatsuki, N., and Asahi, T.,** Purification and properties of two enzymes hydrolyzing synthetic substrates, N-α-benzoyl-D,L-arginine *p*-nitroanilide and leucine *p*-nitroanilide from pea seeds, *Agric. Biol. Chem.,* 43, 315, 1978.

122. **Shutov, A. D., Koroleva, T. N., Khan, N. T. K., and Vaintraub, I. A.**, A protease of germinating vetch seedlings which hydrolyzes native storage proteins, *Dokl. Akad. Nauk SSSR*, 231, 1010, 1976.
123. **Basha, S. M. M. and Cherry, J. P.**, Proteolytic enzyme activity and storage protein degradation in the cotyledons of germinating peanut (*Arachis hypogaea* L.) seeds, *J. Agric. Food Chem.*, 26, 229, 1978.
124. **St. Angelo, A. J., Ory, R. L., and Hansen, H. J.**, Purification of acid proteinase from *Cannabis sativa* L., *Phytochemistry*, 8, 1873, 1969.
125. **Shepard, D. V. and Moore, K. G.**, Purification and properties of a protease from *Lupinus angustifolius* during germination, *Eur. J. Biochem.*, 91, 263, 1978.
126. **Shain, Y. and Mayer, A. M.**, Proteolytic enzymes and endogenous trypsin inhibitor in germinating lettuce seeds, *Physiol. Plant.*, 18, 853, 1965.
127. **Ainouz, I. L. and Ponte, A. L. L.**, Proteolytic activity in *Vigna unguiculata* (L.) Walp. using modified endogenous proteins as substrates, *Plant Physiol.*, 65, 142S, 1980.
128. **Spencer, P. W. and Spencer, R. D.**, Globulin-specific proteolytic activity in germinating pumpkin seeds as detected by a fluorescence assay method, *Plant Physiol.*, 54, 925, 1974.
129. **Van Huystee, R. B.**, Survey of major proteases in germinating peanut seeds by affinity chromatography, *Z. Pflanzenphysiol.*, 89, 51, 1978.
130. **Irving, G. W. and Fontaine, T. D.**, Purification and properties of arachain, a newly discovered proteolytic enzyme of the peanut, *Arch. Biochem.*, 6, 351, 1945.
131. **Shain, Y. and Mayer, A. M.**, Activation of enzymes during germination — trypsin-like enzymes in lettuce, *Phytochemistry*, 7, 1491, 1968.
132. **Shutov, A. D., Koroleva, T. N., and Vaintraub, I. A.**, Participation of proteases of dormant vetch seeds in decomposition of storage proteins during germination, *Fiziol. Rast.*, 25, 735, 1978.
133. **Murray, D. R., Peoples, M. B., and Waters, S. P.**, Proteolysis in the axis of germinating pea. I. Changes in protein degrading enzyme activities of the radicle and primary root, *Planta*, 147, 111, 1979.
134. **Wilson, K. A. and Tan-Wilson, A. L.**, Proteinases involved in the degradation of trypsin inhibitor in germinating mung beans, *Acta Biochim. Pol.*, 30, 139, 1983.
135. **Ofelt, C. N., Smith, A. K., and Mills, J. M.**, Proteases of the soybean, *Cereal Chem.*, 32, 53, 1955.
136. **Bond, H. M. and Bowles, D. J.**, Characterization of soybean endopeptidase activity using exogenous and endogenous substrates, *Plant Physiol.*, 72, 345, 1983.
137. **Wilson, K. A.**, Studies on the Trypsin Inhibitors of the Garden Bean, *Phaseolus vulgaris*, Ph.D. thesis, State University of New York at Buffalo, 1973.
138. **Pusztai, A. and Duncan, I.**, Changes in proteolytic enzyme activities and transformation of nitrogenous compounds in the germinating seeds of kidney bean (*Phaseolus vulgaris*), *Planta*, 96, 317, 1971.
139. **Yatsu, L. Y. and Jacks, T. J.**, Association of lysosomal activity with aleurone grains in plant seeds, *Arch. Biochem. Biophys.*, 124, 466, 1968.
140. **Scharrenberger, C., Oeser, A., and Tolbert, N. E.**, Isolation of protein bodies on sucrose gradients, *Planta*, 104, 185, 1972.
141. **St. Angelo, A. J., Ory, R. L., and Hansen, H. J.**, Properties of a purified proteinase from hempseed, *Phytochemistry*, 9, 1933, 1970.
142. **Wilson, K. A.**, unpublished data, 1973.
143. **Dalkin, K., Marcus, S., and Bowles, D. J.**, Endopeptidase activity in jackbeans and its effect on concanavalin A, *Planta*, 157, 531, 1983.
144. **Korolyova, T. N., Shutov, A. D., and Vaintraub, I. A.**, The action of the proteolytic enzymes of dry vetch seeds on their own reserve proteins, *Plant Sci. Lett.*, 4, 309, 1975.
145. **Harris, N. and Chrispeels, M. J.**, Histochemical and biochemical observations on storage protein metabolism and protein body autolysis in cotyledons of germinating mung beans, *Plant Physiol.*, 56, 292, 1975.
146. **Shinano, S. and Fukushima, K.**, Studies of lotus seed protease. II. Purification and some properties, *Agric. Biol. Chem.*, 33, 1236, 1969.
147. **Guardiola, J. L. and Sutcliffe, J. F.**, Control of protein hydrolysis in the cotyledons of germinating pea (*Pisum sativum* L.) seeds, *Ann. Bot.*, 35, 791, 1971.
148. **Morris, G. F. I., Thurman, D. A., and Boulter, D.**, The extraction and chemical composition of aleurone grains (protein bodies) isolated from seeds of *Vicia faba*, *Phytochemistry*, 9, 1707, 1970.
149. **Tazawa, Y. and Hirokawa, T.**, Uber Pflanzenproteasen. VI. Die proteolytische Aktivitat des Soja-legumellins, *J. Biochem.*, 43, 785, 1956.
150. **Shutov, A. D. and Vaintraub, I. A.**, The degradation of reserve proteins during seed germination. Proteases participating in this process and the role of reserve protein structure, in *Seed Proteins of Dicotyledonous Plants*, Müntz, K., Ed., Akademie-Verlag, Berlin, 1979, 145.
151. **Lichtenfeld, C., Manteufell, R., Müntz, K., Schlesier, B., and Scholz, G.**, Changes in proteolytic activity and protein pattern during germination of field beans (*Vicia faba* L.), in *Seed Proteins of Dicotyledonous Plants*, Muntz, K., Ed., Akademie-Verlag, Berlin, 1979, 163.

152. **Tully, R. E. and Beevers, H.,** Proteases and peptidases of castor bean endosperm. Enzyme characterization and changes during germination, *Plant Physiol.,* 62, 746, 1978.
153. **Feller, U.,** Nitrogen mobilization and proteolytic activities in germinating and maturing bush beans (*Phaseolus vulgaris* L.), *Z. Pflanzenphysiol,* 95, 413, 1979.
154. **Vavreinova, S. and Turkova, J.,** SH-Proteinase from bean *Phaseolus vulgaris* var. Perlicka, *Biochim. Biophys. Acta,* 403, 506, 1975.
155. **Yomo, H. and Varner, J. E.,** Control of the formation of amylase and proteases in the cotyledons of germinating peas, *Plant Physiol.,* 511, 708, 1973.
156. **Shutov, A. D., Do Ngok Lanh, and Vaintraub, I. A.,** Purification and partial characterization of protease B from germinating vetch seeds, *Biokhimiya,* 47, 814, 1982.
157. **Hara, I. and Matsubara, H.,** Pumpkin (*Cucurbita* sp.) seed globulin. VI. Proteolytic activities appearing in germinating cotyledons, *Plant Cell Physiol.,* 21, 233, 1980.
158. **Baumgartner, B., Tokuyasu, K. T., and Chrispeels, M. J.,** Localization of vicilin peptidohydrolase in the cotyledons of mung bean seedlings by immunofluorescence microscopy, *J. Cell Biol.,* 79, 10, 1978.
159. **Chrispeels, M. J., Baumgartner, B., and Harris, N.,** Regulation of reserve protein metabolism in the cotyledons of mung bean seedlings, *Proc. Natl. Acad. Sci. U.S.A.,* 73, 3168, 1976.
160. **Petra, P. H.,** Bovine procarboxypeptidase and carboxypeptidase A, *Methods Enzymol.,* 19, 460, 1970.
161. **Folk, J. E.,** Carboxypeptidase B (porcine pancreas), *Methods Enzymol.,* 19, 504, 1970.
162. **Ihle, J. N. and Dure, L. S.,** The developmental biochemistry of cottonseed embryogenesis and germination. I. Purification and properties of a carboxypeptidase from germinating cotyledons, *J. Biol. Chem.,* 247, 5034, 1972.
163. **Wells, J. R. E.,** Purification and properties of a proteolytic enzyme from french beans, *Biochem. J.,* 97, 228, 1965.
164. **Nishimura, M. and Beevers, H.,** Hydrolases in vacuoles from castor bean endosperm, *Plant Physiol.,* 62, 44, 1978.
165. **Mikola, J.,** Activities of various peptidases in cotyledons of germinating peanut *(Arachis hypogaea), Physiol. Plant.,* 36, 255, 1976.
166. **Kabota, Y., Shoji, S., Yamanaka, T., and Yamato, M.,** Carboxypeptidases from germinating soybeans. I. Purification and properties of two carboxypeptidases, *Yakugaku Zasshi,* 96, 639, 1976.
167. **Van der Wilden, W., Herman, E. M., and Chrispeels, M. J.,** Protein bodies of mung bean cotyledons as autophagic organelles, *Proc. Natl. Acad. Sci. U.S.A.,* 77, 428, 1980.
168. **Ihle, J. and Dure, L. S.,** The developmental biochemistry of cottonseed embryogenesis and germination. III. Regulation of the biosynthesis of enzymes utilized in germination, *J. Biol. Chem.,* 247, 5048, 1972.
169. **Elleman, T. C.,** Aminopeptidases of pea, *Biochem. J.,* 141, 113, 1974.
170. **Caldwell, J. B. and Sparrow, L. G.,** Partial purification and characterization of two peptide hydrolases from pea seeds, *Plant Physiol.,* 57, 795, 1976.
171. **Cameron, E. C. and Mazelis, M.,** A nonproteolytic "trypsin-like" enzyme. Purification and properties of arachain, *Plant Physiol.,* 48, 278, 1971.
172. **Mainguy, P. N. R., Van Huystee, R. B., and Hayden, D. B.,** Form and action of a protease in cotyledons of germinating peanut seeds, *Can. J. Bot.,* 50, 2189, 1972.
173. **Catsimpoolas, N., Funk, S. K., Wang, J., and Kenney, J.,** Isoelectric fractionation and some properties of a protease from soybean seeds, *J. Sci. Food Agric.,* 22, 79, 1971.
174. **Shutov, A. D., Nguen Than Uen, Fridman, S. A., and Vaintraub, I. A.,** An investigation of some arylamidases from vetch seeds, *Biokhimiya,* 40, 553, 1975.
175. **Zemchik, E. I., Shutov, A. D., and Vaintraub, I. A.,** Partial purification and characterization of arylamidase from vetch seeds, *Biokhimiya,* 38, 964, 1973.
176. **Zemchik, E. I., Fam Min Than, Shutov, A. D., and Vaintraub, I. A.,** Isolation and further characterization of benzoyl-D,L-arginine p-nitroanilide-hydrolyzing enzyme (BAPAase) from vetch seedlings, *Biokhimiya,* 40, 746, 1975.
177. **Emtserva, I. B. and Belozerskii, M. A.,** N-$\alpha$-benzoyl-D,L-arginine *p*-nitroanilidase from buckwheat seeds. Properties and substrate specificity, *Biokhimiya,* 42, 726, 1977.
178. **Ashton, F. M. and Dahmen, W. J.,** A partial purification and characterization of two aminopeptidases from *Cucurbita maxima* cotyledons, *Phytochemistry,* 6, 641, 1967.
179. **Ashton, F. M. and Dahmen, W. J.,** Purification and characterization of a dipeptidase from *Cucurbita maxima* cotyledons, *Phytochemistry,* 6, 1215, 1967.
180. **Wilson, K. A.,** unpublished data, 1983.
181. **Sze, H. and Ashton, F. M.,** Dipeptidase development in cotyledons of *Cucurbita maxima* during germination, *Phytochemistry,* 10, 2935, 1971.
182. **Wilson, K. A. and Chen, J. C.,** Amino acid sequence of mung bean trypsin inhibitor and its modified forms appearing during germination, *Plant Physiol.,* 71, 341, 1983.
183. **Vazquez, D.,** Inhibitors of protein synthesis, *FEBS Lett.,* 40, S63, 1974.

184. **Hojima, Y., Moriya, H., Moriwaki, C., and Ryan, C. A.,** Metalloenzyme inhibitor from kidney beans. Partial purification and characterization, *Plant Physiol.*, 63, 562, 1979.
185. **Royer, A., Miege, M. D., Grange, A., Miege, J., and Mascherpa, J. M.,** Inhibiteurs anti-trypsine et activites proteolytiques des albumines de graine de *Vigna unguiculata*, *Planta*, 119, 1, 1974.
186. **Royer, A.,** Activities proteolytiques et anti-trypsine des graines de *Vigna unguiculata:* repartition et interaction, *Phytochemistry*, 14, 915, 1975.
187. **Gennis, L. S. and Cantor, C. R.,** Double-headed protease inhibitors from black-eyed peas. I. Purification of two new protease inhibitors and the endogenous protease by affinity chromatography, *J. Biol. Chem.*, 251, 734, 1976.
188. **Walsh, K. A. and Wilcox, P. E.,** Serine proteases, *Methods Enzymol.*, 19, 31, 1970.
189. **Hara, I., Wada, K., and Matsubara, H.,** Pumpkin (*Cucurbita* sp.) seed globulin. II. Alterations during germination, *Plant Cell Physiol.*, 17, 815, 1976.
190. **Hara, I., Ohmiya, M., and Matsubara, H.,** Pumpkin (*Cucurbita* sp.) seed globulin. III. Comparison of subunit structures among seed globulins of various Cucurbita species and characterization of peptide components, *Plant Cell Physiol.*, 19, 237, 1978.
191. **Ohmiya, M., Hara, I., and Matsubara, H.,** Pumpkin (*Cucurbita* sp.) seed globulin. IV. Terminal sequences of the acidic and basic peptide chains and identification of a pyroglutaminyl peptide chain, *Plant Cell Physiol.*, 21, 157, 1980.
192. **Hara, I. and Matsubara, H.,** Degradation process of pumpkin seed globulin during germination, *Plant Physiol.*, 65, 103S, 1980.
193. **Hara, I. and Matsubara, H.,** Pumpkin (*Cucurbita* sp.) seed globulin. VII. Immunofluorescent study on protein bodies in ungerminated and germinating cotyledon cells, *Plant Cell Physiol.*, 21, 247, 1980.
194. **Hara-Nishimura, I., Nishimura, M., Matsubara, H., and Akazawa, T.,** Suborganellar localization of proteinase catalyzing the limited hydrolysis of pumpkin globulin, *Plant Physiol.*, 70, 699, 1982.
195. **Reilly, C. C., O'Kennedy, B. T., Titus, J. S., and Splittoesser, W. E.,** The solubilization and degradation of pumpkin seed globulin during germination, *Plant Cell Physiol.*, 19, 1235, 1978.
196. **Shutov, A. D. and Vaintraub, I. A.,** Chromatographic isolation and some properties of the legumin and vicilin of vetch, *Biokhimiya*, 31, 726, 1966.
197. **Shutov, A. D., Bulmaga, V. P., Boldt, E. K., and Vaintraub, I. A.,** Investigation of the modification of reserve proteins of vetch seeds during germination and limited proteolysis, *Biokhimiya*, 46, 841, 1981.
198. **Tan-Wilson, A. L. and Wilson, K. A.,** unpublished data, 1983.
199. **Kern, R. and Chrispeels, M. J.,** Influence of the axis on the enzymes of protein and amide metabolism in the cotyledons of mung bean, *Plant Physiol.*, 62, 815, 1978.
200. **Chrispeels, M. J. and Baumgartner, B.,** Serological evidence confirming the assignment of *Phaseolus aureus* and *P. mungo* to the genus *Vigna*, *Phytochemistry*, 17, 125, 1978.
201. **Ericson, M. C. and Chrispeels, M. J.,** Isolation and characterization of glucosamine-containing storage glycoproteins from the cotyledons of *Phaseolus aureus*, *Plant Physiol.*, 52, 98, 1973.
202. **Alpi, A. and Beevers, H.,** Effects of leupeptin on proteinase and germination of castor beans, *Plant Physiol.*, 68, 851, 1981.
203. **Yoshikawa, M., Kiyohara, T., Iwasaki, T., and Yoshida, I.,** Modification of proteinase inhibitor II in adzuki beans during germination, *Agric. Biol. Chem.*, 43, 1989, 1979.
204. **Hartl, P. and Wilson, K. A.,** unpublished data, 1984.
205. **Freed, R. C. and Ryan, D. S.,** Changes in Kunitz trypsin inhibitor during germination of soybeans, *J. Food Sci.*, 43, 1316, 1978.
206. **Odani, S., Koide, T., and Ikenaka, T.,** The amino acid sequence of Bowman-Birk soybean trypsin proteinase inhibitor, *Proc. Jpn. Acad.*, 47, 621, 1971.
207. **Lis, H., Sharon, N., and Katchalski, E.,** Soybean hemagglutinin, a plant glycoprotein. I. Isolation of a glycopeptide, *J. Biol. Chem.*, 241, 684, 1966.

Chapter 3

# PROTEOLYTIC ENZYMES IN RELATION TO LEAF SENESCENCE

## Urs Feller

### TABLE OF CONTENTS

| | | |
|---|---|---|
| I. | Introduction | 50 |
| II. | Changes in Leaf N-Metabolism During Senescence | 50 |
| | A. Natural Senescence | 50 |
| | B. Artificial Senescence | 51 |
| III. | Peptide Hydrolase Activities in Senescing Leaves | 51 |
| | A. Exopeptidases | 51 |
| |     1. Aminopeptidases | 51 |
| |     2. Carboxypeptidases | 52 |
| | B. Endopeptidases | 53 |
| |     1. Changes in Total Activity | 53 |
| |     2. Changes in the Pattern of Endopeptidases | 56 |
| IV. | Regulation of Proteolysis | 57 |
| | A. Levels of Peptide Hydrolase Activities Present | 57 |
| | B. Susceptibility of Substrate Proteins | 58 |
| | C. Compartmentation | 59 |
| | D. Effect of pH and Low-Molecular-Weight Compounds | 60 |
| V. | Conclusions | 62 |
| Acknowledgments | | 63 |
| References | | 63 |

## I. INTRODUCTION

Leaf metabolism changes considerably with the onset of senescence.[1-3] The assimilation of inorganic nutrients ceases and previously assimilated nutrients are remobilized. The activities of enzymes involved in the assimilation of carbon,[4,5] nitrogen,[6-8] and sulfur[9] decrease. Proteins,[7-10] nucleic acids,[10] chlorophylls,[11-13] carotenoids,[13] and membrane lipids[14] are degraded. Low-molecular-weight compounds formed (e.g., amino acids) can be translocated to other parts of the plant. Nutrients mobilized by the breakdown of essential leaf compounds become available for sinks within the same plant.[15,16] Leaf senescence is, therefore, important in the nutrient economy of higher plants.

Senescence is not a passive decay of the leaf.[1-3] It is a well-organized and regulated process. The importance of a proper senescence control becomes evident from the fact that a leaf is able to initiate its own disassembly. Different metabolic processes are lost gradually during leaf senescence. Respiration remains active longer than photosynthesis.[17,18] Some enzymes (e.g., several hydrolases) reach maximal activities during senescence and may contribute to rapid nutrient mobilization.

The proteins degraded during leaf senescence[19] are different from the storage proteins mobilized during seed germination.[20] The substrates for proteolytic enzymes in leaves are mainly enzyme and membrane proteins. A large portion of the leaf proteins is present in the chloroplasts.[21] Ribulose bisphosphate carboxylase (RuBPCase) can contribute more than 50% to the soluble leaf proteins in $C_3$-plants.[21-23] The proteins hydrolyzed and their subcellular localization are different for germinating seeds and for senescing leaves.

A rapid net degradation of leaf proteins occurs during senescence, but proteolysis is not restricted to senescence. It must be kept in mind that peptide bonds are hydrolyzed throughout leaf development. Proteolytic processing takes place when proteins are transported from the cytosol of eukaryotic cells into the endoplasmic reticulum, into mitochondria, or into chloroplasts.[24-27] The processing enzymes in eukaryotic cells were found to be highly specific for the cleavage of a defined peptide from the precursor protein (see Chapter 4, Volume II). A partially purified chelator-sensitive protease from the matrix of yeast mitochondria did not hydrolyze either other peptide bonds of the processed protein or nonmitochondrial proteins.[26]

Since leaf proteins are under turnover, protein synthesis and proteolysis occur even then, when no net changes in the protein contents are detectable.[28] The recycling of amino acids from degraded proteins requires a complete hydrolysis to free amino acids by peptide hydrolases present in the leaf. A half-life of 5 to 8 days was determined for RuBPCase in corn leaves.[28] Different half-lives were found for soluble proteins and for membrane proteins in barley.[29] The half-lives can be affected by environmental factors.[29]

This chapter focuses on protein degradation during leaf senescence and does not cover protein processing and protein turnover in nonsenescing leaves. The question arises as to how the rapid mobilization of leaf proteins could be controlled. Possible regulation mechanisms at the cellular level are discussed here in addition to the changes in the nitrogen metabolism and in the pattern of peptide hydrolases during leaf senescence.

## II. CHANGES IN LEAF N-METABOLISM DURING SENESCENCE

### A. Natural Senescence

Natural leaf senescence is defined here as the senescence behavior of leaves attached to intact plants not subjected to environmental stress. Naturally senescing leaves lose their ability to assimilate nitrate, and a net degradation of nitrogen containing macromolecules can be observed. In general, the low-molecular-weight compounds formed are exported rapidly and do not accumulate within the leaf.[30-33] An intact long-distance transport system is a prerequisite for the redistribution of nitrogen and other nutrients during natural senescence.

The pattern of enzymes involved in nitrogen metabolism changes in this stage of development. The activities of nitrate reductase,[6,8] glutamine synthetase,[7,8] and glutamate synthase[8] decrease, while glutamate dehydrogenase[8,34] and endopeptidases[30-33,35,36] remain active longer and often reach maximal activities during senescence. It was proposed that glutamate dehydrogenase could be involved rather in the catabolism than in the formation of glutamate.[37,38] Though glutamine synthetase activity decreases, the remaining activity was found to be sufficient for the reassimilation of ammonia produced by the breakdown of amino acids.[7]

Different time courses during senescence were observed for enzyme forms present inside and outside the chloroplasts. A more rapid inactivation was detected for glutamine synthetase in the chloroplasts than for the cytosolic form.[39,40] Similar results were obtained for aminotransferases in leaves of *Lolium temulentum* during artificial senescence.[41] The activities of the forms present in the chloroplasts decreased earlier than the cytosolic forms.

### B. Artificial Senescence

Senescence processes were frequently investigated in detached shoots, detached leaves, or leaf segments.[21,38,41-45] Hormonal interactions between different organs and the export of compounds from the leaf are no longer possible after detachment.[3,46,47] Sugars[44,48] and amino acids[38,45,48,49] can accumulate in the detached leaf. The high concentrations of low-molecular-weight compounds are likely to affect the leaf metabolism and to interfere with the breakdown of macromolecules. An accumulation of amino acids was also detected in leaves of dark-treated intact cereal plants,[48,50] and increased sugar contents were measured in corn[33] and wheat[44] leaves after sink removal. It was proposed that the increased carbohydrate concentration could cause senescence.[33,44]

The metabolic changes and the pattern of peptide hydrolase activities observed during artificial senescence are not necessarily identical with those in intact plants, allowing a rapid export of breakdown products.[45] Artificial senescence is a helpful tool for detail studies, but caution is recommended for the extrapolation of such results to the situation in intact plants (see Chapter 8, Section II.B.2).

## III. PEPTIDE HYDROLASE ACTIVITIES IN SENESCING LEAVES

### A. Exopeptidases
*1. Aminopeptidases*

Aminopeptidases have been detected in different forms in a series of plant species.[51-55] The various forms were separated by gel electrophoresis and by ion exchange chromatography.[51-55] They differ in their specificities for the N-terminal amino acids of the substrate and in their inhibition properties. The pH optima for enzymes of this group are in the neutral or slightly alkaline region (see Chapter 5, Volume I).

No major increases in aminopeptidase activities were detected during leaf senescence. In some cases, these enzyme activities remained relatively constant or increased slightly during the early phase of nitrogen mobilization.[53,56] In general, aminopeptidase activities were found to be high in developing and mature plant parts.[57-59] In senescing leaves of maize,[36] *Lolium temulentum*,[38] wheat,[18,31] dwarf beans,[58] and a series of other annual and perennial plants,[32] aminopeptidase activities decreased in a similar manner to leaf protein content. The time courses for nitrogen mobilization and peptide hydrolase activities are shown in Figure 1 for the three uppermost leaves of field-grown wheat.

In barley and wheat leaves, a large portion of the aminopeptidase activity was localized in the chloroplasts.[60,61] Only traces of aminopeptidase activity were detected in the vacuoles of barley leaves.[60]

The time course of aminopeptidase activities suggests that enzymes of this group may not

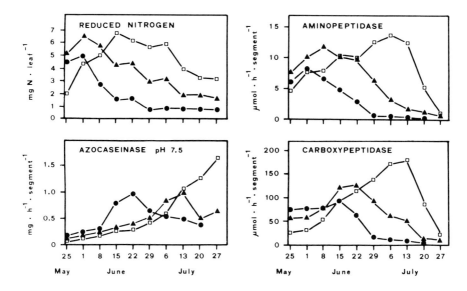

FIGURE 1. Changes in the nitrogen content and in peptide hydrolase activities of wheat leaves during natural senescence. Flag (□), second (▲), and third (●) leaves from top were harvested and analyzed at intervals of 1 week. The samples for the reduced nitrogen contained the full leaf spreads, but not the leaf sheaths. The enzyme activities were measured in segments containing the middle 50% in length of the leaf spread. (From Feller, U. and Erismann, K. H., Z. Pflanzenphysiol., 90, 235, 1978. With permission.)

play a major role in the protein mobilization from senescing leaves. The high activities in developing and mature leaves support the hypothesis that aminopeptidases could be involved in protein turnover or modification, rather than in senescence. However, the functions of aminopeptidases are not yet sufficiently elucidated to allow final conclusions (see Chapter 5, Volume I).

2. Carboxypeptidases

Carboxypeptidases have been isolated and characterized from leaves of citrus,[62-64] french bean,[65,66] and tomato plants.[67-69] Maximal activities with artificial peptide substrates (e.g., N-carbobenzoxy dipeptides) were observed in the range between pH 5 and 6.5. Leaf carboxypeptidases liberate a wide range of C-terminal amino acids from proteins or oligopeptides. In several studies, plant carboxypeptidases were found to be sensitive to diisopropylfluorophosphate or to phenylmethylsulfonyl fluoride, indicating that a serine residue may be in or near the active center[36,57,66-69] (see Chapter 5, Volume I).

A several-fold increase in carboxypeptidase activity was observed in all leaves of tomato plants after wounding one or two lower leaves.[67-69] The contents of two proteinase inhibitors increased simultaneously with carboxypeptidase activity, but aminopeptidase and endopeptidase activities were not affected by wounding.[69] Therefore, wounding caused a specific increase in carboxypeptidase and not a general increase in peptide hydrolase activities. The increased carboxypeptidase activities in the upper leaves of these tomato plants are not necessarily related to senescence, since this enzyme may be involved in processes other than the remobilization of leaf proteins.[69]

Relatively low carboxypeptidase activities were measured in expanding wheat leaves, while highest activities were detected in fully expanded mature leaves. In general, this enzyme activity decreases in senescing leaves (Figure 1). However, carboxypeptidases reach maximal levels later than aminopeptidases in leaves of corn, wheat, and several other plant species. During senescence, carboxypeptidase was found to remain active longer than aminopeptidase. The thermal stability in vitro is better for carboxypeptidase.[70] The two groups of exopeptidases differ considerably in the time courses during development and senescence, in the pH- and temperature-optima, in the inhibition properties as well as in their subcellular distribution. Carboxypeptidase activity was localized to a large extent in the vacuoles of higher plant cells, while aminopeptidase activity was detected mainly outside the vacuoles.[60,61,68]

Although carboxypeptidase activities decrease during senescence in some plant species, it appears possible that this enzyme contributes to the rapid degradation of leaf proteins. No evidence was presented so far for a control function of carboxypeptidase in the senescence mobilization of leaf proteins. Possibly this enzyme contributes to the further hydrolysis of peptides produced by endopeptidases.

## B. Endopeptidases
### 1. Changes in Total Activity

Endopeptidases from various plant species have been reviewed by Frith and Dalling[71] and by Ryan and Walker-Simmons.[72] The enzymes characterized so far are sensitive to sulfhydryl group inhibitors,[72] but whether they can be correctly categorized as cysteine-proteinases is not yet clear (see Chapters 1 and 2, Volume I). The pH optimum of the main activity against hemoglobin, casein, or RuBPCase was often found in the region of 4 to 7.[71-74] The pH optima for proteolytic enzymes present in extracts from senescing wheat leaves depended on the substrate used.[50] The maximal activity against hemoglobin was observed around pH 4, while casein was hydrolyzed most rapidly around pH 5.5.[50] Broad optima around pH 5 were observed in assays with RuBPCase or with cytochrome $c$ as substrates.[50] It is, therefore, difficult to compare results from different groups using unequal assay procedures.

Plant endopeptidases not belonging to the cysteine-proteinases were found with pH optima in the acidic (pH 3 to 5) and in the alkaline region.[72,75,76] Apparently the serine endopeptidases known from microorganisms and animals (e.g., trypsin, chymotrypsin) are not present in high activities in leaves.[72]

Measurements of endopeptidase activities with crude extract are difficult because of possible interactions with exopeptidases.[77] Proteins with only one C- and one N-terminal amino acid may be poor substrates for the exopeptidases, but the concentration of their substrates (C- and N-terminal amino acids) can be increased by endopeptidase activity. If exopeptidases are present in the extract, a further hydrolysis of peptides produced by endopeptidases cannot be ruled out.[36,77]

The levels of endopeptidase activities in mature leaves and their time courses during leaf development and senescence differ considerably between various groups of annual and perennial plants.[32] It is necessary to discuss the changes in endopeptidase activities during senescence separately for the different plant groups. Annual crop plants with monocarpic senescence are of special interest in this context, because the importance of a rapid nutrient remobilization from senescing leaves is obvious. In these plants a close relationship between leaf senescence and seed filling exists. On the other hand, nutrient remobilization from leaves of perennial plants is not closely related to seed maturation and may be controlled in a different manner.

Leaves of Gramineae were used in several laboratories to study protein mobilization during senescence. Endopeptidase activities are low in expanding and mature leaves. Marked increases in these enzyme activities were observed during natural[18,23,35,36] or artificial[21,38,42] senescence. Endopeptidases reach maximal activities at a time when aminopeptidase and carboxypeptidase activities are already declining (Figure 1). The rise in endopeptidase activities starts at the tip of the individual leaves and parallels the visible senescence symptoms.[18,36] The changes in the peptide hydrolase activities later reach the basal parts of the leaf and finally the leaf sheath. The enzyme pattern changes sequentially in the different leaves of field-grown wheat[18] and corn.[36] A good correlation was found for the activities of hemoglobin-degrading enzymes and the actual nitrogen loss from senescing parts of maturing wheat plants.[35] Often endopeptidase activities reached maximal levels when the bulk of leaf nitrogen was already exported to other plant parts.[18,35,36] The time courses and the broad range of substrates hydrolyzed are consistent with the hypothesis that endopeptidases play a major role in the mobilization of proteins from senescing leaves. Relatively high endopeptidase activities were observed also in naturally senescing parts of wheat ears.[31,35,78,79] In glumes (consisting of glume, lemma, and palea), in the outer pericarp and in the cross cells, endopeptidase activities peaked during maturation, while these enzyme activities remained low in the embryo and in the endosperm.[78,79] Nitrogen was mobilized sequentially from the different senescing layers of maturing wheat kernels. A high endopeptidase activity and a decreasing nitrogen content was first detected in the outer pericarp and later in the chlorophyll-containing cross cells.[79] In dark-incubated, detached leaves of corn,[49] oat,[80] and in leaves of detached wheat shoots,[45] a rapid net protein breakdown was observed without a major increase in total endopeptidase activity. These results suggest that the high endopeptidase activities observed during natural senescence of cereal leaves may not necessarily be a prerequisite for nitrogen mobilization. Considerable endopeptidase activities were already present at the beginning of the experiments. Under these conditions, the net protein breakdown was not initiated by an increase in total endopeptidase activity. The results mentioned above are consistent with the assumption that endopeptidases are important in the mobilization of proteins from senescing plant parts. Since the level of total endopeptidase activity appears unlikely to control protein mobilization, the regulation mechanisms for the net protein breakdown remain open.

In different legume species, proteolytic enzymes were detected, which were stimulated by sulfhydryl reagents and inhibited by p-hydroxymercuribenzoate or N-ethylmaleimide.[81-84] Ragster and Chrispeels[77,84,85] were able to show that the sulfhydryl-dependent enzymes from soybean leaves were endopeptidases, while the activity inhibited by phenylmethylsulfonyl fluoride was likely to be due to a carboxypeptidase. In clover,[81] pea,[83] and soybean[84] leaves, major proteolytic activities were observed in the range of pH 4 to 7. Legume endopeptidases with rather high pH optima (in the range of pH 9 to 10) were reported by several groups.[75,76,85] Azocoll-digesting proteinases with pH optima around 9 increased during leaf expansion and maturation when protein and chlorophyll contents increased.[85] During senescence, these enzyme activities declined. Furthermore, it was shown with inhibitors that the azocoll-digesting proteinases are unlikely to play a major role in the protein breakdown occurring in crude leaf extracts.[85] The major azocoll-digesting enzyme extracted from senescing soybean leaves (azocollase A) was strongly inhibited by EDTA, while azocollase B (the predominant form in young leaves) was inhibited by p-hydroxymercuribenzoate. The proteolysis observed in crude extracts at pH 9 was not affected by these inhibitors, suggesting only a minor role for the azocollases in the degradation of major leaf proteins.[85] In general, no good correlation was found between the levels of proteolytic activities and protein mobilization from senescing legume leaves.[58,75-77,83-85]

A clear increase in total proteolytic activity during senescence of legume leaves was observed only in a few cases.[32,86] Storey and Beevers[83] reported increased proteolytic activities during senescence in pods, but not in leaves of pea plants. The contradictory results on changes in peptide hydrolase activities during legume leaf senescence suggest that different factors may influence the time courses for the total proteolytic activity (e.g., involvement of different peptide hydrolases, nutrition status of the plants, stress). Control of protein mobilization during legume leaf senescence by the total endopeptidase activity appears unlikely.

A sulfhydryl-stimulated proteolytic activity with a pH optimum around 5 was extracted from leaves of Solanaceae.[32,87,88] The major activity detected in extracts from tobacco leaves was identified as an endopeptidase, since mainly peptides and not free amino acids were produced during incubation.[88] Peptide hydrolase activities declined in tobacco leaves during senescence.[88,89] Since the protein content declined more rapidly than the absolute enzyme activity, the specific activity increased during senescence.[88] Hochkeppel[90] isolated an endopeptidase from senescing tobacco leaves, which hydrolyzed hydrophobic membrane proteins. The molecular weight of this endopeptidase was only 9,000 to 10,000, and maximal activity was observed around pH 5. No enzyme activity was detectable in green leaves, but a precipitate was formed with antibodies against the endopeptidase. The author concluded that either the enzyme was present in an inactive form, being activated during senescence, or the enzyme activity present in green leaf extracts was not sufficient to be detected with the methods used. The question remained open, whether the enzyme was synthesized *de novo* or an inactive form was activated during senescence.[90] Large quantities of serine proteinase inhibitors were extracted from leaves of Solanaceae.[91,92] These inhibitors affect serine endopeptidases from microorganisms and animals, but no inhibition of endogenous peptide hydrolases from these leaves was detected.[69,92] These inhibitors may be involved, rather, in the protection against plant pests than in the regulation of proteolysis in the leaves.[99]

Only a few reports concern peptide hydrolases in senescing leaves of trees.[32,94,95] Highest proteolytic activities were detected in apple leaves during senescence, but the protein content decreased before these enzyme activities increased.[94] Proteolytic activities were maximal after most of the leaf proteins were already degraded.[94,95] RuBPCase was hydrolyzed most rapidly in the pH range of 4.5 to 5 by a sulfhydryl-sensitive endopeptidase.[95] The endopeptidase activities extracted from senescing apple leaves, larch, or pine needles were relatively low compared to leaves of annual crop plants.[32] On the other hand, nitrogen was mobilized more slowly in trees than in cereals or legumes.[94,95]

Some plants contain extremely high endopeptidase activities.[72,96-101] Enzymes from these sources were used to study the mechanism of action, the inhibition by specific and unspecific reagents, and to identify the preferentially cleaved peptide bonds.[72] Several cysteine proteinases were isolated from the latex of Caricaceae (e.g., *Carica papaya*)[96] and of Moraceae (e.g., *Ficus carica*),.[99,100] Pineapple fruits, stems, and leaves contain high endopeptidase activities.[97,98,102] Pineapple leaf endopeptidases were found to be vacuolar enzymes.[102] Very high proteolytic activities were recently reported for vegetable samples of ginger and green asparagus and for fruit samples of kiwi, but the enzymes contributing to these high activities are not yet well characterized.[101] The endopeptidase activities in the species mentioned above are generally already high in developing and mature plant parts. It appears unlikely that these extremely high endopeptidase activities are important for the protein mobilization during senescence. The physiological role of the elevated endopeptidase levels in certain plant species is still open (see Chapter 4, Volume I).

Endopeptidases were found to be mainly located in the vacuoles of pineapple,[102] wheat,[61,103,104] barley,[60] and pea[105] leaf cells. Although large portions of these enzymes were found in the vacuoles, there are indications that relatively unspecific endopeptidases (not identical with processing enzymes cleaving specific peptide bonds) are present also in other cell compartments.[84,104,106,107] If endopeptidases and their substrate proteins are in the same compartment, the regulation of the protein breakdown becomes the central problem.

The properties of the main proteolytic activity detected in senescing leaves differed from those in extracts from germinating seeds of the same species.[36,57,58,70,72] The pH optimum, the temperature optimum, and the thermal stability were different for the azocasein hydrolyzing activities in crude extracts from bean leaves and cotyledons of germinating bean seeds.[58,70] The pH profiles and the inhibition properties of the caseolytic activity in endosperm extracts of germinating corn seeds differed from those of roots and leaves.[36,57] These results suggest that different endopeptidases or different endopeptidase patterns are present in senescing leaves and in germinating seeds (see Chapters 1 and 2).

## 2. Changes in the Pattern of Endopeptidases

A change in the pattern of endopeptidase activities in leaves of Gramineae was suggested by the observation that the pH optimum of crude extracts shifted during senescence from the slightly acidic to the neutral or slightly alkaline pH range.[6,32,33,36,38,42,108] Therefore, the time courses for proteolytic activities around pH 5 were considerably different from those at or above pH 7. The neutral activity was often still increasing when the slightly acidic activity was already declining. Thomas[38] mentioned the possibility of a modification (limited self-hydrolysis?) of the peptide hydrolase protein altering its pH optimum. However, there is no evidence to support this speculation. The shift of the pH optimum in crude leaf extracts was observed by several groups during natural[32,36,108] or artificial[38,42] senescence. Often the neutral activity peaked late in senescence, when most of the leaf proteins were already hydrolyzed.[6,32,33,36,108]

Several endopeptidases from leaves of Gramineae were separated by ion exchange chromatography.[109-113] Differences for the separated enzyme activities were observed in their pH optima,[109,113] response to inhibitors,[109-111] substrate specificities,[111,114] and thermal stabilities.[110] Endopeptidases extracted from wheat[115] or barley[111] leaves hydrolyzed proteins from animal sources (hemoglobin, azocasein) as well as the major leaf protein RuBPCase. These endopeptidases are, therefore, rather unspecific. Some endopeptidase activity was found to be associated with purified RuBPCase.[107,111,114] This activity was very low and not detectable with standard assay procedures, but SDS-PAGE revealed significant degradation of RuBPCase. The large and small subunits of RuBPCase were hydrolyzed differently by the separated endopeptidases.[114] Alterations in the proportions of the different activities were observed besides the increase in total endopeptidase activity during leaf senescence.[112,113] All activities detected in senescing leaves were present also in mature leaves. From the data available so far for cereal leaves, it appears unlikely that the rapid protein mobilization in senescing leaves is initiated by an increase in a senescence-specific endopeptidase activity.

In naturally senescing dwarf bean leaves, the optimal pH for the degradation of azocasein shifted from the alkaline to the acidic pH range.[76] This is in contrast to the observations on cereal leaves, where the pH optima for extracts of senescing leaves were always higher than those for extracts from mature leaves.[32,36,38,42] Major differences in the senescence behavior and in the pattern of proteolytic enzymes of legumes and cereals are likely, since the time courses for the total endopeptidase activity are also different in the two plant groups.[58,76,77,83] Various endopeptidase activities from bean leaves were separated by column chromatography[116] and by isoelectric focusing.[76] The shift in the pH optimum of crude extracts was paralleled by changes in the endopeptidase pattern.[76] More bands were visible on zymograms of senescing than of nonsenescing leaves, but the authors could not rule out that the enzymes were already present in low amounts in mature leaves. The increasing (or newly formed) endopeptidase activities hydrolyzed gelatin most rapidly in the range of pH 3.5 to 5.[76] In cowpea, the substrate specificity of the proteolytic activity in crude extracts changed during leaf development and senescence, suggesting changes in the pattern of peptide hydrolases.[117] The endopeptidases from soybean leaves were separated on columns, assayed with various substrates, and further characterized by Chrispeels and co-workers.[77,84,85] Two enzymes degrading azocoll (azocollase A and B) showed maximal activities in the alkaline pH range (around pH 9).[85] Azocollase A was sensitive to EDTA, while azocollase B was inhibited by *p*-hydroxymercuribenzoate. Azocollase A liberated free amino groups from casein or denatured leaf protein. The authors concluded that azocollase A and B are two distinct enzymes, rather than different forms of the same enzyme. Azocollase B was predominant in young leaves, while in senescing leaves the total activity was mainly due to azocollase A. The azocoll-digesting enzymes increased during leaf maturation and decreased during leaf senescence. A major role for these enzymes in leaf senescence appeared unlikely.[85] More recently it was shown by the same group that in leaves of *Phaseolus vulgaris* the azocoll-degrading activity was present in the cell wall.[118] Three hemoglobin-digesting endopeptidases were found besides the two azocollases in soybean leaves.[77] These enzymes were most active around pH 4 with hemoglobin or RuBPCase as substrates. Hemoglobinase I not binding to DEAE-cellulose at pH 7.5 was more active in young leaves than the two other hemoglobin-degrading enzymes. In mature and senescing leaves, hemoglobinase III (the most anionic of the three activities) was predominant. Hemoglobinase III remained active in vitro only in the presence of sulfhydryl reagents and was strongly inhibited by N-ethylmaleimide and p-hydroxymercuribenzoate. The other two hemoglobin-degrading endopeptidases were much less sensitive to these inhibitors. All three enzymes were present in all stages of leaf development. During senescence these activities declined in soybean leaves.[77]

Although changes in the endopeptidase pattern were observed during senescence of leaves from Gramineae and Leguminosae, senescence-specific endopeptidases have not been unequivocally identified so far. The proportions of the different activities changed, but the enzymes were detected in most of the species investigated in considerable activities throughout leaf development. Therefore, other mechanisms than the formation of a "senescence" endopeptidase must be considered for the regulation of protein mobilization from leaves. The physiological roles of the various endopeptidases, therefore, remain to be elucidated (see Chapter 6, Volume I).

## IV. REGULATION OF PROTEOLYSIS

Possible regulation mechanisms for intracellular protein breakdown were discussed by Holzer and Heinrich[119] and more recently by Hershko and Ciechanover.[120] The degradation of a particular protein depends on the peptide hydrolase activities present in the same compartment, on the susceptibility of the protein to proteolytic attack, and on the biochemical conditions in this compartment (e.g., pH, concentrations of metabolites). Several factors may be involved in the control of proteolysis in cells of senescing leaves. Some of these factors are discussed here in more detail, but the different possibilities are speculative as long as data remains insufficient to consolidate a general concept.

### A. Levels of Peptide Hydrolase Activities Present

Proteolysis can be accelerated by enhanced levels of peptide hydrolases. Increased amounts of proteolytic enzymes in a cell compartment can be due to *de novo* synthesis of the enzyme protein,[21,42] activation of a zymogen,[90] disappearance of inhibitors,[121] or changes in the compartmentation.[122-124] Senescing leaves contain enzyme inactivating factors, which are most likely proteolytic enzymes (Figure 2). The inactivating factor was heat sensitive, precipitated by ammonium sulfate and excluded by Sephadex® G-25.[125] The inactivation of enzymes was delayed after addition of other proteins to the extract, e.g., casein or heat-denatured leaf extract. A similar inactivation of enzymes was observed after addition of papain, trypsin, or chymotrypsin to leaf extracts.[40,126] These results illustrate the attack of a particular enzyme in vitro, but it should be considered that in intact leaf cells these enzymes may be separated by membranes from the main proteolytic activities.[60,61,103-105] The time courses for peptide hydrolase activities in whole leaves or leaf segments do not allow conclusions on the activities present in different cell compartments. The reduction in protein breakdown after application of protein synthesis inhibitors suggests, but cannot prove, that a *de novo* synthesis of peptide hydrolases is involved in the rapid protein breakdown in these leaves. Further information on the subcellular distribution of proteolytic activities throughout leaf development and senescence is needed to decide whether or not enhanced peptide hydrolase activities are a major control factor for the protein loss in a particular compartment.

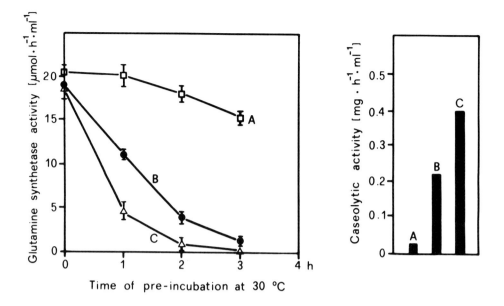

FIGURE 2. Caseolytic activities and inactivation of glutamine synthetase in wheat leaf extracts. A — extract from young leaves mixed with buffer (1:2); B — extract from young leaves mixed with buffer and with extract from senescing leaves (1:1:1); C — extract from young leaves mixed with extract from senescing leaves (1:2). The means ±SD are shown for the glutamine synthetase activity. The bars for the proteolytic activity represent the means of duplicate measurements at the beginning of the preincubation. (From Streit, L. and Feller, U., *Experientia*, 38, 1176, 1982. With permission.)

## B. Susceptibility of Substrate Proteins

Various plant and animal proteins are degraded by leaf proteolytic enzymes.[50,83,115,127] All peaks detected after chromatography of wheat leaf extract on DEAE-cellulose degraded hemoglobin and RuBPCase, indicating that these proteolytic enzymes are not necessarily highly specific for leaf proteins.[115] However, the degradation rates vary in a wide range between substrates.[50,83] Different susceptibilities to plant and animal endopeptidases were observed for a series of plant enzymes.[126,128]

Nitrate reductase is known as a highly susceptible enzyme, while glutamate dehydrogenase is very stable under the same conditions.[126,128,129] It was shown that inactivation of nitrate reductase is a rather complex process, which is not necessarily due to proteolysis.[130] The involvement of proteolytic enzymes was suggested for some systems, where the formation of breakdown products paralleled nitrate reductase inactivation.[130,131] The inactivating factor extracted from barley leaves was inhibited by leupeptin, an inhibitor of some endopeptidases.[131] If proteolytic enzymes and their substrates are present in the same compartment, the relative susceptibilities of the different proteins can cause different degradation rates within this compartment.

Unequal inactivation rates were observed for two forms of glutamine synthetase (Figure 3). Glutamine synthetase 2 (the form present in chloroplasts) was more rapidly inactivated than glutamine synthetase 1 (cytosolic enzyme) in an extract from senescing wheat leaves containing both forms.[40] Only minor inactivation was observed at 2°C, while most of the activity of glutamine synthetase 2 was lost during 2 hr at 30°C. These results are consistent with the observation that glutamine synthetase 2 is less stable at high temperatures than the cytosolic form.[132] During senescence, the cytosolic enzyme remains active longer than the chloroplast enzyme.[39,40] Both, the resistance to proteolytic attack and the subcellular location, may contribute to the different time courses for two forms.

Other proteins present in the same compartment can protect a particular enzyme, since

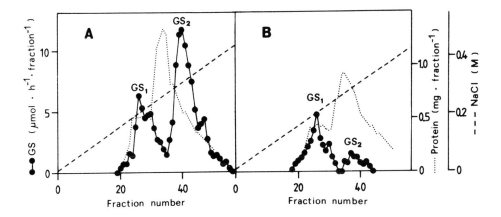

FIGURE 3. Stability of two forms of glutamine synthetase (GS) in extract from senescing wheat leaves. The extract was preincubated in stoppered test tubes for 2 hr at 4°C (A) or at 30°C (B) prior to column chromatography. (From Streit, L. and Feller, U., *Physiol. Vég.*, 21, 103, 1983. With permission.)

they may represent competitive substrates for peptide hydrolases. In a similar manner, enzymes in leaf extracts can be protected in vitro by the addition of casein or denatured leaf proteins.[125] Furthermore, the susceptibility of a protein can be altered by covalent modifications[133] or by interactions with low-molecular-weight compounds.[125] It appears possible that in some cases modifications of the substrate proteins preceding the cleavage of peptide-bonds initiate the breakdown of the protein.

### C. Compartmentation

The spatial separation of proteolytic enzymes and their substrates is a possible control mechanism for the degradation of leaf proteins. A particular protein is protected from proteolysis as long as the degradative enzymes are located in other cell compartments. Carboxypeptidases and endopeptidases with pH optima in the range of 4 to 7 were detected by several investigators, mainly in the vacuoles.[60,61,89,102-105] Alkaline endopeptidase activity with a pH optimum around 9 was found to be located in the extracellular fluid of bean leaves.[118] Only a very small portion of the leaf proteins is present in the vacuole and in the cell wall, while the bulk of leaf proteins is located inside the plasmalemma and outside the vacuole.[102,103] The chloroplasts contain a high percentage of the total leaf proteins.[21-23] Therefore, most of the leaf proteins are separated by membranes from the major proteolytic activity. It was proposed that chloroplasts are degraded sequentially in the vacuoles of senescing wheat leaves[124] (see Chapter 8). Another mechanism was suggested by Cooke et al.[122] for deuterium oxide-induced protein degradation in *Lemna minor*. According to these authors, the properties of the tonoplast may be altered allowing release of vacuolar peptide hydrolases into the cytoplasm, or, alternatively, facilitating the entry of selected proteins into the vacuole. It is not clear yet to which extent vacuolar enzymes are involved in the mobilization of proteins from senescing leaves. If proteins from the chloroplasts, from the mitochondria, or from the cytosol should be degraded by vacuolar enzymes, the crucial question would be how the proteolytic enzymes get in contact with the substrate proteins.

It was suggested by several investigators that chloroplast constituents may be degraded within the chloroplasts.[10,12,84,134,135] Proteolytic activities were also detected in isolated chloroplasts.[61,84,103,135] The presence of degradative enzymes in the chloroplasts could allow the breakdown of proteins inside these organelles without prior changes in the compartmentation. If proteolytic enzymes and their substrates should be located in the same compartment in vivo, the question arises how the net breakdown could be regulated.

Membrane proteins are in a special situation. They are interacting with membrane lipids. Interesting results were obtained from a nonyellowing mutant of *Festuca pratensis*.[112,136,137] Soluble chloroplast proteins decreased in the mutant in a similar manner as in a normally yellowing genotype.[136] Membrane polypeptides and chlorophyll, on the other hand, were not degraded in the mutant. From the proteolytic enzymes present in leaf extracts, it was concluded that the accessibility of the membrane proteins, rather than the pattern of peptide hydrolases, was different in the nonyellowing mutant.[112] Membrane proteins may represent a special case of compartmentation. They may be protected from proteolysis as long as they are embedded in the hydrophobic membrane. Fatty acids were considered to play a major role in the degradation of membrane proteins.[137] The data obtained in experiments with *Festuca pratensis* suggest the hydrolysis of thylakoid membranes and of soluble proteins inside the chloroplasts.[134,136] From the behavior of the nonyellowing mutant, it appears unlikely that chloroplasts are degraded sequentially in the central vacuole.[137] These results support the hypothesis that net protein breakdown occurs within the chloroplasts of senescing leaves (see Chapter 8).

The contributions of the different peptide hydrolases to the mobilization of leaf proteins during senescence is still unclear. It should be considered that proteolytic enzymes located in different compartments may be involved in the complete hydrolysis of a particular protein. Peptides produced by endopeptidases may be further hydrolyzed to free amino acids in a different cell compartment. Under this aspect, it appears possible that the vacuole is involved in the hydrolysis of peptides produced from protein breakdown in other compartments. The initial, well-ordered phase of protein mobilization may differ from the final stage of senescence when compartmentation may be affected by membrane disruption.

### D. Effect of pH and Low-Molecular-Weight Compounds

The pH value, inorganic ions, or metabolites may influence proteolysis by interactions with either the proteolytic enzymes or the substrate proteins. The pH optimum for the degradation of a protein depends not only on the properties of the peptide hydrolases, but also on those of the substrate protein.[50] Therefore, changes in the pH value of a compartment are likely to accelerate or delay proteolysis. The degradation of various proteins may be affected differently by a shift in pH.

Kaur-Sawhney et al.[138] reported that polyamines are potent inhibitors of senescence. Much lower azocoll-degrading activities were extracted from oat leaf segments treated for 48 hr with polyamines than in control segments floated on buffer alone. The marked increase in proteolytic activities observed in control segments during artificial senescence did not occur in the presence of polyamines. Furthermore, in vitro proteolytic enzyme activity was inhibited by spermine or spermidine, indicating that proteolysis may be affected directly by these compounds. The authors concluded from experiments with pretreated enzyme or substrate that the inhibition of proteolysis by polyamines was mainly due to interactions with the proteolytic enzymes and not with the substrate. A pretreatment of the enzyme extract was as effective as a direct addition of polyamines to the enzyme assay, while a pretreatment of the in vitro-substrate (azocoll) was ineffective. Considerable changes in the polyamine content were observed during artificial senescence of oat leaf segments.[139] Therefore, polyamines should be considered as possible regulators of proteolysis in vivo (see Chapter 7, Volume I).

Anionic detergents and fatty acids can accelerate the degradation of proteins.[135,137,140,141] Since the degradation of three yeast enzymes was affected differently by the addition of oleate, it appears likely that the effect of fatty acids was primarily due to interactions with the substrate proteins rather than with the proteolytic enzymes. The pH profiles for the degradation of RuBPCase by endogenous peptide hydrolases of purified barley chloroplasts differed in the presence and absence of sodium dodecyl sulfate.[135] The contents of free fatty

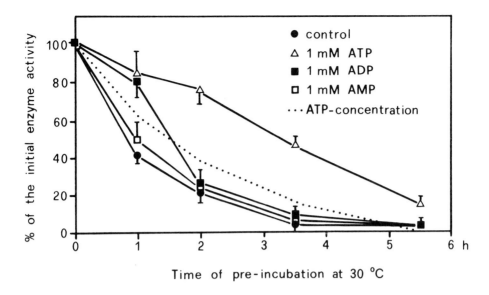

FIGURE 4. Effect of adenylates on glutamine synthetase inactivation. Extract from young wheat leaves was mixed 1:1 with extract from senescing leaves before the effectors were added from stock solutions. The dotted line indicates the actual ATP concentration in the treatments with 1 mM ATP. The symbols represent the means of five separate preincubations. The standard deviations are shown only on one side for clarity. (From Streit, L. and Feller, U., *Experientia*, 38, 1176, 1982. With permission.)

acids can change in vivo during senescence.[143] The degradation of membrane lipids and the concentrations of free fatty acids may be involved in the control of proteolysis in chloroplasts as suggested by Thomas.[137]

The susceptibility of enzymes to proteolytic attack can be altered by substrates, cofactors, stimulators, or inhibitors.[119,125,142] The stabilization of glutamine synthetase in wheat leaf extracts is illustrated in Figure 4. This enzyme was protected in the presence of ATP, while ADP or AMP had little or no effect.[125] The initial stabilization in the presence of ADP may be due to the formation of ATP by adenylate kinase activity present in the crude extract. In contrast to glutamine synthetase, nitrate reductase inactivation was not affected by ATP, ADP, or AMP, but was influenced by pyridine nucleotides.[125] Glucose-6-phosphate dehydrogenase in endosperm extracts of germinating wheat seeds was stabilized in the presence of inorganic phosphate or pyridine nucleotides.[142] NADPH and NADP were better protectants than NADH or NAD. A similar protection of this enzyme by pyridine nucleotides was observed if the inactivation was accelerated after addition of trypsin or chymotrypsin. The large subunit of RuBPCase in isolated chloroplasts of *Festuca pratensis* was markedly stabilized by ATP when the pH of the incubation medium was above 7.[143] ATP may stabilize in this case indirectly. The results obtained from different systems suggest a fine control of proteolysis by metabolic changes. Lamb tubulin treated with chymotrypsin remained active when GTP was present in the incubation medium, while GDP caused only a minor effect.[144] It was shown that tubulin was hydrolyzed in presence and absence of GTP. Two fragments were produced. The authors concluded that GTP was not affecting the cleavage of tubulin by chymotrypsin, but stabilized directly or indirectly the two fragments in an active conformation. In this case, GTP could not prevent proteolysis, but protected from conformational changes and from the activity loss as a consequence of the tubulin cleavage. Since different proteins are stabilized by other low-molecular-weight compounds, partially selective effects on some proteins appear possible. The data mentioned above support the hypothesis that an accelerated protein breakdown can be initiated without changes in the activities or in the compartmentation of proteolytic enzymes. A better knowledge of the factors influencing the susceptibilities of individual proteins could be a key to understanding the mechanisms involved in the fine control of proteolysis in leaves.

ATP-dependent proteolytic sytems were detected in microorganisms,[145] animals,[120,146-149] and recently also in spinach leaves[150] and pea chloroplasts.[151,152] In some cases, ubiquitin (a small polypeptide) was required for the ATP-dependent proteolysis,[120,147,148,153-155] while in other systems ubiquitin has no effect.[120,145,146,156] For the ATP-dependent protein breakdown in reticulocytes, it was proposed that conjugates of ubiquitin with proteins could be formed prior to the hydrolysis of peptide bonds[148,153-155] or that ubiquitin may be involved in releasing a peptide hydrolase from its endogenous inhibitor.[149] It was observed that oxidized, incomplete, or abnormal proteins were rapidly degraded by the ATP-dependent proteolytic system from erythrocytes, while unmodified complete proteins were much more stable.[148,149,157] The hydrolysis of ATP and proteins in a linked fashion was reported for protease La present in *Escherichia coli*.[145] The mechanisms and the functions of ATP-dependent proteolytic systems are still hypothetical. The involvement of ATP-dependent proteolysis in senescing leaves remains to be elucidated. ATP was found not to be required for the degradation of the herbicide-binding protein (32-kdalton protein from thylakoid membranes) by a membrane-bound endopeptidase.[158] ATP may stimulate proteolysis by activation of ATP-dependent proteolytic systems in some cases[145-157] or delay the breakdown of some proteins (e.g., glutamine synthetase) by interactions with the substrate proteins.[125]

## V. CONCLUSIONS

Typical changes in peptide hydrolase activities during senescence were observed in leaves of a series of plants, but an increase in proteolytic activities was not always correlated with the onset of the net protein breakdown. No unequivocal evidence for a senescence-specific peptide hydrolase could be found so far. It appears, therefore, likely that the net protein breakdown during senescence is not initiated by an increase in total proteolytic activity or in the activity of a specific peptide hydrolase. Other mechanisms than the formation of proteolytic activity must be considered for the regulation of the net protein breakdown. One could argue that leaf proteins are under turnover, and the net breakdown could be caused simply by the cessation of protein synthesis. There is evidence for a more rapid protein breakdown in senescing leaves than in mature leaves, indicating an enhanced proteolysis during senescence. A major gap in our present knowledge of leaf senescence concerns the mechanisms involved in the regulation of proteolysis on the cellular and subcellular level.

The bulk of the leaf proteins is outside the vacuole, while endopeptidase and carboxypeptidase activities were mainly localized in the vacuole.[60,61,102-105] The question arises where and by which peptide hydrolases the leaf proteins are degraded. It appears possible that several compartments contribute to the complete hydrolysis. Further investigations are required to answer the question as to where plastidial, mitochondrial, peroxisomal, or cytosolic proteins are hydrolyzed. It remains open, whether or not different proteolytic systems are involved in the breakdown of proteins from various cell compartments.

A series of possible fine control mechanisms were suggested. Membrane proteins may be protected as long as they interact with membrane lipids.[112] The degradation of these proteins may, therefore, depend on lipid metabolism. Low-molecular-weight compounds can specifically affect the susceptibility of some proteins.[125,140,142,158] Such interactions could allow a specific regulation for the breakdown of certain proteins. Our knowledge on the susceptibility of different proteins to proteolysis and on regulatory effects of pH, inorganic ions, or metabolites is still rudimentary. Perhaps proteolysis is controlled in vivo to a large extent by changes in the susceptibilities of substrate proteins by covalent modification or by interactions with low-molecular-weight compounds. Such control mechanisms could be much more sensitive to metabolic changes and more specific than effects on peptide hydrolases. The susceptibility of a particular enzyme could be altered immediately after a change in the metabolism. The physiological role of ATP-stimulated proteolysis in leaves should be further investigated.[150-152] It is a challenge to elucidate the role of the different peptide hydrolases and the mechanisms involved in the control of proteolysis in order to understand the remobilization of leaf proteins during senescence and the regulation of this process.

## ACKNOWLEDGMENTS

I thank Dr. J. C. Rutter and Dr. J. Fuhrer for improving the English of the manuscript, M. Baumgartner for preparing the photographs of the figures, and I. Feller for organizing the reference list and typing the manuscript. The literature search and recent unpublished work was supported in part by Swiss National Science Foundation, Project 3.067-0.81.

## REFERENCES

1. **Stoddart, J. L. and Thomas, H.**, Leaf senescence, in *Encyclopedia of Plant Physiology*, New Series, Vol. 14A, Boulter, D. and Parthier, B., Eds., Springer-Verlag, Berlin, 1982, 592.
2. **Woolhouse, H. W.**, Leaf senescence, in *The Molecular Biology of Plant Development*, Smith, H. and Grierson, D., Eds., University of California Press, Berkeley, 1982, 256.
3. **Noodén, L. D.**, Senescence in the whole plant, in *Senescence in Plants*, Thimann, K. V., Ed., CRC Press, Boca Raton, Fla., 1980, 219.
4. **Hall, N. P., Keys, A. J., and Merrett, M. J.**, Ribulose-1,5-diphosphate carboxylase protein during flag leaf senescence, *J. Exp. Bot.*, 29, 31, 1978.
5. **Thomas, S. M., Hall, N. P., and Merrett, M. J.**, Ribulose 1,5-bisphosphate carboxylase/oxygenase activity and photorespiration during the aging of flag leaves of wheat, *J. Exp. Bot.*, 29, 1161, 1978.
6. **Reed, A. J., Below, F. E., and Hageman, R. H.**, Grain protein accumulation and the relationship between leaf nitrate reductase and protease activities during grain development in maize (*Zea mays* L.). I. Variation between genotypes, *Plant Physiol.*, 66, 164, 1980.
7. **Storey, R. and Beevers, L.**, Enzymology of glutamine metabolism related to senescence and seed development in the pea (*Pisum sativum* L.), *Plant Physiol.*, 61, 494, 1978.
8. **Streit, L. and Feller, U.**, Changing activities of nitrogen-assimilating enzymes during growth and senescence of dwarf beans (*Phaseolus vulgaris* L.), *Z. Pflanzenphysiol.*, 108, 273, 1982.
9. **Schmutz, D. and Brunold, C.**, Regulation of sulfate assimilation in plants. XIII. Assimilatory sulfate reduction during ontogenesis of primary leaves of *Phaseolus vulgaris* L., *Plant Physiol.*, 70, 524, 1982.
10. **Misra, A. N. and Biswal, U. C.**, Changes in the content of plastid macromolecules during aging of attached and detached leaves, and of isolated chloroplasts of wheat seedlings, *Photosynthetica*, 16, 22, 1982.
11. **Martinoia, E., Dalling, M. J., and Matile, Ph.**, Catabolism of chlorophyll: demonstration of chloroplast-localized peroxidative and oxidative activities, *Z. Pflanzenphysiol*, 107, 269, 1982.
12. **Martinoia, E., Heck, U., Dalling, M. J., and Matile, Ph.**, Changes in chloroplast number and chloroplast constituents in senescing barley leaves, *Biochem. Physiol. Pflanzen*, 178, 147, 1983.
13. **Matile, P. and Martinoia, E.**, Catabolism of carotenoids: involvement of peroxidase?, *Plant Cell Rep.*, 1, 244, 1982.
14. **Harwood, J. L., Jones, A. V. H. M., and Thomas, H.**, Leaf senescence in a non-yellowing mutant of *Festuca pratensis*, *Planta*, 156, 152, 1982.
15. **Pate, J. S. and Layzell, D. B.**, Carbon and nitrogen partitioning in the whole plant — a thesis based on empirical modeling, in *Nitrogen and Carbon Metabolism*, Bewley, J. D., Ed., Martinus Nijhoff/Junk, The Hague, 1981, 94.
16. **Simpson, R. J., Lambers, H., and Dalling, M. J.**, Nitrogen redistribution during grain growth in wheat (*Triticum aestivum* L.). IV. Development of a quantitative model of the translocation of nitrogen to the grain, *Plant Physiol.*, 71, 7, 1983.
17. **Wittenbach, V. A.**, Induced senescence of intact wheat seedlings and its reversibility, *Plant Physiol.*, 59, 1039, 1977.
18. **Feller, U. and Erismann, K. H.**, Veränderungen des Gaswechsels und der Aktivitäten proteolytischer Enzyme während der Seneszenz von Weizenblättern (*Triticum aestivum* L.), *Z. Pflanzenphysiol.*, 90, 235, 1978.

19. **Huffaker, R. C. and Peterson, L. W.**, Protein turnover in plants and possible means of its regulation, *Annu. Rev. Plant Physiol.*, 25, 363, 1974.
20. **Ashton, F. M.**, Mobilization of storage proteins of seeds, *Annu. Rev. Plant Physiol.*, 27, 95, 1976.
21. **Peterson, L. W. and Huffaker, R. C.**, Loss of ribulose 1,5-diphosphate carboxylase and increase in proteolytic activity during senescence of detached primary barley leaves, *Plant Physiol.*, 55, 1009, 1975.
22. **Gordon, K. H. J., Peoples, M. B., and Murray, D. R.**, Ageing-linked changes in photosynthetic capacity and in fraction I protein content of the first leaf of pea *Pisum sativum* L., *New Phytol.*, 81, 35, 1978.
23. **Wittenbach, V. A.**, Ribulose bisphosphate carboxylase and proteolytic activity in wheat leaves from anthesis through senescence, *Plant Physiol.*, 64, 884, 1979.
24. **Ellis, R. J.**, Protein transport across membranes: an introduction, *Biochem. Soc. Symp.*, 46, 223, 1981.
25. **Schatz, G.**, How are proteins imported into mitochondria?, *Cell*, 32, 316, 1983.
26. **McAda, P. C. and Douglas, M. G.**, A neutral metallo endoprotease involved in the processing of an $F_1$-ATPase subunit precursor in mitochondria, *J. Biol. Chem.*, 257, 3177, 1982.
27. **Ellis, R. J.**, Chloroplast protein synthesis: principles and problems, in *Subcellular Biochemistry*, Vol. 9, Roody, D. B., Ed., Plenum Publishing, 1983, 237.
28. **Simpson, E., Cooke, R. J., and Davies, D. D.**, Measurement of protein degradation in leaves of *Zea mays* using ($^3$H)acetic anhydride and tritiated water, *Plant Physiol.*, 67, 1214, 1981.
29. **Dungey, N. O. and Davies, D. D.**, Protein turnover in the attached leaves of non-stressed and stressed barley seedlings, *Planta*, 154, 435, 1982.
30. **Feller, U.**, Effect of changed source/sink relations on proteolytic activities and on nitrogen mobilization in field-grown wheat (*Triticum aestivum* L.), *Plant Cell Physiol.*, 20, 1577, 1979.
31. **Waters, S. P., Peoples, M. B., Simpson, R. J., and Dalling, M. J.**, Nitrogen redistribution during grain growth in wheat (*Triticum aestivum* L.). I. Peptide hydrolase activity and protein breakdown in the flag leaf, glumes and stem, *Planta*, 148, 422, 1980.
32. **Keist, M.**, Vergleich der extrahierbaren Aktivitaeten von Peptidhydrolasen in Blaettern verschiedener Pflanzengruppen vor und waehrend der Seneszenz, Lizentiatsarbeit, University of Berne, Switzerland, 1980.
33. **Christensen, L. E., Below, F. E., and Hageman, R. H.**, The effects of ear removal on senescence and metabolism of maize, *Plant Physiol.*, 68, 1180, 1981.
34. **Lauriere, C., Weisman, N., and Daussant, J.**, Glutamate dehydrogenase in the first leaf of wheat. II. De novo synthesis upon darkness stress and senescence, *Physiol. Plant*, 52, 151, 1981.
35. **Dalling, M. J., Boland, G., and Wilson, J. H.**, Relation between acid proteinase activity and redistribution of nitrogen during grain development in wheat, *Aust. J. Plant Physiol.*, 3, 721, 1976.
36. **Feller, U. K., Soong, T.-S. T., and Hageman, R. H.**, Leaf proteolytic activities and senescence during grain development of field-grown corn (*Zea mays* L.), *Plant Physiol.*, 59, 290, 1977.
37. **Miflin, B. J. and Lea, P. J.**, Amino acid metabolism, *Annu. Rev. Plant Physiol.*, 28, 299, 1977.
38. **Thomas, H.**, Enzymes of nitrogen mobilization in detached leaves of *Lolium temulentum* during senescence, *Planta*, 142, 161, 1978.
39. **Mann, A. F., Fentem, P. A., and Stewart, G. R.**, Tissue localization of barley (*Hordeum vulgare*) glutamine synthetase isoenzymes, *FEBS Lett.*, 110, 265, 1980.
40. **Streit, L. and Feller, U.**, Changing activities and different resistance to proteolytic activity of two forms of glutamine synthetase in wheat leaves during senescence, *Physiol. Vég.*, 21, 103, 1983.
41. **Thomas, H.**, Regulation of alanine aminotransferase in leaves of *Lolium temulentum* during senescence, *Z. Pflanzenphysiol.*, 74, 208, 1975.
42. **Martin, C. and Thimann, K. V.**, The role of protein synthesis in the senescence of leaves. I. The formation of protease, *Plant Physiol.*, 49, 64, 1972.
43. **Tetley, R. M. and Thimann, K. V.**, The metabolism of oat leaves during senescence. I. Respiration, carbohydrate metabolism, and the action of cytokinins, *Plant Physiol.*, 54, 294, 1974.
44. **Lazan, H. B., Barlow, E. W. R., and Brady, C. J.**, The significance of vascular connection in regulating senescence of the detached flag leaf of wheat, *J. Exp. Bot.*, 34, 726, 1983.
45. **Feller, U.**, Senescence and proteolytic activities in detached leaves and detached shoots of wheat, *Physiol. Vég.*, 21, 93, 1983.
46. **Hsia, C. P. and Kao, C. H.**, The importance of roots in regulating the senescence of soybean primary leaves, *Physiol. Plant*, 43, 385, 1978.
47. **Malik, N. S. A. and Berrie, A. M. M.**, The role of roots in shoot senescence of peas (*Pisum sativum* L.), *Z. Pflanzenphysiol.*, 100, 79, 1980.
48. **Thimann, K. V., Tetley, R. R., and Van Thanh, T.**, The metabolism of oat leaves during senescence. II. Senescence in leaves attached to the plant, *Plant Physiol.*, 54, 859, 1974.
49. **Soong, T.-S. T., Feller, U. K., and Hageman, R. H.**, Changes in activities of proteolytic enzymes during senescence of detached corn (*Zea Mays* L.) leaves as function of physiological age, *Plant Physiol.*, 59 (Suppl.), 112, 1977.

50. **Wittenbach, V. A.**, Breakdown of ribulose bisphosphate carboxylase and change in proteolytic activity during dark-induced senescence of wheat seedlings, *Plant Physiol.*, 62, 604, 1978.
51. **Collier, M. D. and Murray, D. R.**, Leucyl β-naphthylamidase activities in developing seeds and seedlings of *Pisum sativum* L., *Aust. J. Plant Physiol.*, 4, 571, 1977.
52. **Waters, S. P. and Dalling, M. J.**, Distribution and characteristics of aminoacyl β-naphthylamidase activities in wheat seedlings, *Aust. J. Plant Physiol.*, 6, 595, 1979.
53. **Tazaki, K. and Ishikura, N.**, Multiple forms of aminopeptidase in *Euonymus* leaves, *Plant Cell Physiol.*, 24, 1263, 1983.
54. **Scandalios, J. G. and Espiritu, L. G.**, Mutant aminopeptidases of *Pisum sativum*. I. Developmental genetics and chemical characteristics, *Molec. Gen. Genet.*, 105, 101, 1969.
55. **Ott, L. A. and Scandalios, J. G.**, Genetically defined peptidases of maize. I. Biochemical characterization of allelic and nonallelic forms, *Biochem. Genet.*, 14, 619, 1976.
56. **Sopanen, T. and Laurière, C.**, Activities of various peptidases in the first leaf of wheat, *Physiol. Plant*, 36, 251, 1976.
57. **Feller, U., Soong, T.-S. T., and Hageman, R. H.**, Patterns of proteolytic enzyme activities in different tissues of germinating corn (*Zea mays* L.), *Planta*, 140, 155, 1978.
58. **Feller, U.**, Nitrogen mobilization and proteolytic activities in germinating and maturing bush beans (*Phaseolus vulgaris* L.), *Z. Pflanzenphysiol*, 95, 413, 1979.
59. **Dahlhelm, H. and Schober, H.**, Beziehungen zwischen proteolytischer Aktivität und Proteinabbau in höheren Pflanzen. II. Der Einfluss von exogen applizierten Phytohormonen auf die proteolytische Aktivität und den Protein-Turnover in *Pisum sativum*, *Biochem. Physiol. Pflanzen*, 176, 770, 1981.
60. **Heck, U., Martinoia, E., and Matile, P.**, Subcellular localization of acid proteinase in barley mesophyll protoplasts, *Planta*, 151, 198, 1981.
61. **Waters, S. P., Noble, E. R., and Dalling, M. J.**, Intracellular localization of peptide hydrolases in wheat (*Triticum aestivum* L.) leaves, *Plant Physiol.*, 69, 575, 1982.
62. **Zuber, H. and Matile, Ph.**, Acid carboxypeptidases: their occurrence in plants, intracellular distribution and possible function, *Z. Naturforsch.*, 23b, 663, 1968.
63. **Sprössler, B., Heilmann, H. D., Grampp, E., and Uhlig, H.**, Eigenschaften der Carboxypeptidase C aus Orangenblättern, *Hoppe-Seyler's Z. Physiol. Chem.*, 352, 1524, 1971.
64. **Zuber, H.**, Carboxypeptidase C, in *Methods Enzymol.*, 45 (Part B), 561, 1976.
65. **Wells, J. R. E.**, Purification and properties of a proteolytic enzyme from french beans, *Biochem. J.*, 97, 228, 1965.
66. **Carey, W. F. and Wells, J. R. E.**, Phaseolain: a plant carboxypeptidase of unique specificity, *J. Biol. Chem.*, 247, 5573, 1972.
67. **Walker-Simmons, M. and Ryan, C. A.**, Wound-induced peptidase activity in tomato leaves, *Biochem. Biophys. Res. Commun.*, 74, 411, 1977.
68. **Walker-Simmons, M. and Ryan, C. A.**, Immunological identification of proteinase inhibitors I and II in isolated tomato leaf vacuoles, *Plant Physiol.*, 60, 61, 1977.
69. **Walker-Simmons, M. and Ryan, C. A.**, Isolation and properties of carboxypeptidase from leaves of wounded tomato plants, *Phytochemistry*, 19, 43, 1980.
70. **Feller, U.**, In vitro stability and inactivation of peptide hydrolases extracted from *Phaseolus vulgaris* L., *Plant Cell Physiol.*, 22, 1095, 1981.
71. **Frith, G. J. T. and Dalling, M. J.**, The role of peptide hydrolases in leaf senescence, in *Senescence in Plants*, Thimann, K. V., Ed., CRC Press, Boca Raton, Fla., 1980, 117.
72. **Ryan, C. A. and Walker-Simmons, M.**, Plant proteinases, in *The Biochemistry of Plants*, Vol. 6, Marcus, A., Ed., Academic Press, New York, 1981, 321.
73. **Pike, C. S. and Briggs, W. R.**, Partial purification and characterization of a phytochrome-degrading neutral protease from etiolated oat shoots, *Plant Physiol.*, 49, 521, 1972.
74. **Peoples, M. B. and Dalling, M. J.**, Degradation of ribulose-1,5-bisphosphate carboxylase by proteolytic enzymes from crude extracts of wheat leaves, *Planta*, 138, 153, 1978.
75. **Racusen, D. and Foote, M.**, An endopeptidase of bean leaves, *Can. J. Bot.*, 48, 1017, 1970.
76. **Weckenmann, D. and Martin, P.**, Changes in the pattern of endopeptidases during senescence of bush bean leaves (*Phaseolus vulgaris* L.), *Z. Pflanzenphysiol.*, 104, 103, 1981.
77. **Ragster, L. E. and Chrispeels, M. J.**, Hemoglobin-digesting acid proteinases in soybean leaves. Characteristics and changes during leaf maturation and senescence, *Plant Physiol.*, 67, 110, 1981.
78. **Kruger, J. E.**, Changes in the levels of proteolytic enzymes from hard red spring wheat during growth and maturation, *Cereal Chem.*, 50, 122, 1973.
79. **Feller, U.**, Changes in nitrogen contents and in proteolytic activities in different parts of field-grown wheat ears (*Triticum aestivum* L.) during maturation, *Plant Cell Physiol.*, 19, 1489, 1978.
80. **Van Loon, L. C., Haverkort, A. J., and Lokhorst, G. J.**, Changes in Protease Activity During Leaf Growth and Senescence, Proc. FESPP Inaugural Meeting at Edinburgh, 1978, 544.

81. **Brady, C. J.**, The leaf protease of *Trifolium repens*, *Biochem. J.*, 78, 631, 1961.
82. **Vavreinova, S. and Turkova, J.**, SH-proteinase from bean *Phaseolus vulgaris* var. *Perlicka*, *Biochim. Biophys. Acta*, 403, 506, 1975.
83. **Storey, R. and Beevers, L.**, Proteolytic activity in relationship to senescence and cotyledonary development in *Pisum sativum* L., *Planta*, 137, 37, 1977.
84. **Ragster, L. E. and Chrispeels, M. J.**, Autodigestion in crude extracts of soybean leaves and isolated chloroplasts as a measure of proteolytic activity, *Plant Physiol.*, 67, 104, 1981.
85. **Ragster, L. and Chrispeels, M. J.**, Azocoll-digesting proteinases in soybean leaves. Characteristics and changes during leaf maturation and senescence, *Plant Physiol.*, 64, 857, 1979.
86. **Wittenbach, V. A., Ackerson, R. C., Giaquinta, R. T., and Hebert, R. R.**, Changes in photosynthesis, ribulose bisphosphate carboxylase, proteolytic activity, and ultrastructure of soybean leaves during senescence, *Crop Sci.*, 20, 225, 1980.
87. **Tracey, M. V.**, Leaf protease of tobacco and other plants, *Biochem. J.*, 42, 281, 1948.
88. **Anderson, J. W. and Rowan, K. S.**, Activity of peptidase in tobacco-leaf tissue in relation to senescence, *Biochem. J.*, 97, 741, 1965.
89. **Balz, H. P.**, Intrazelluläre Lokalisation und Funktion von hydrolytischen Enzymen bei Tabak, *Planta*, 70, 207, 1966.
90. **Hochkeppel, H.-K.**, Isolierung einer Endopeptidase aus alternden Tabakblättern und ihre Beziehung zum Vergilben, *Z. Pflanzenphysiol*, 69, 329, 1973.
91. **Walker-Simmons, M. and Ryan, C. A.**, Wound-induced accumulation of trypsin inhibitor activities in plant leaves, *Plant Physiol.*, 59, 437, 1977.
92. **Santarius, K. and Belitz, H.-D.**, Proteinase activity in potato plants, *Planta*, 141, 145, 1978.
93. **Ryan, C. A.**, Proteinase inhibitor, in *The Biochemistry of Plants*, Vol. 6, Marcus, A., Ed., Academic Press, New York, 1981, 351.
94. **Spencer, P. W. and Titus, J. S.**, Biochemical and enzymatic changes in apple leaf tissue during autumnal senescence, *Plant Physiol.*, 49, 746, 1972.
95. **Kang, S.-M., Matsui, H., and Titus, J. S.**, Characteristics and activity changes of proteolytic enzymes in apple leaves during autumnal senescence, *Plant Physiol.*, 70, 1367, 1982.
96. **Lynn, K. R.**, A purification and some properties of two proteases from papaya latex, *Biochim. Biophys. Acta*, 569, 193, 1979.
97. **Murachi, T.**, Bromelain enzymes, in *Methods Enzymol.*, 45, (Part B), 475, 1976.
98. **Daley, L. S. and Vines, H. M.**, Pineapple (*Ananas comosus* L., merr.) leaf proteinase, *Plant Sci. Lett.*, 11, 59, 1978.
99. **Sgarbieri, V. C., Gupte, S. M., Kramer, D. E., and Whitaker, J. R.**, Ficus enzymes. I. Separation of the proteolytic enzymes of *Ficus carica* and *Ficus glabrata* latices, *J. Biol. Chem.*, 239, 2170, 1964.
100. **Kramer, D. E. and Whitaker, J. R.**, Ficus enzymes. II. Properties of the proteolytic enzymes from the latex of *Ficus carica* variety kadota, *J. Biol. Chem.*, 239, 2178, 1964.
101. **Yamaguchi, T., Yamashita, Y., Takeda, I., and Kiso, H.**, Proteolytic enzymes in green asparagus, kiwi fruit and miut: occurrence and partial characterization, *Agric. Biol. Chem.*, 46, 1983, 1982.
102. **Boller, T. and Kende, H.**, Hydrolytic enzymes in the central vacuole of plant cells, *Plant Physiol.*, 63, 1123, 1979.
103. **Lin, W. and Wittenbach, V. A.**, Subcellular localization of proteases in wheat and corn mesophyll protoplasts, *Plant Physiol.*, 67, 969, 1981.
104. **Wanger, G. J., Mulready, P., and Cutt, J.**, Vacuole/extravacuole distribution of soluble protease in *Hippeastrum* petal and *Triticum* leaf protoplasts, *Plant Physiol.*, 68, 1081, 1981.
105. **Noble, E. R. and Dalling, M. J.**, Intracellular localization of acid peptide hydrolases and several other acid hydrolases in the leaf of pea (*Pisum sativum* L.), *Aust. J. Plant Physiol.*, 9, 353, 1982.
106. **Peoples, M. B. and Dalling, M. J.**, Intracellular localization of acid peptide hydrolases in wheat leaves, *Plant Physiol.*, 63 (Suppl.), 159, 1979.
107. **Thomas, H. and Huffaker, R. C.**, Hydrolysis of radioactively-labelled ribulose-1,5-bisphosphate carboxylase by an endopeptidase from the primary leaf of barley seedlings, *Plant Sci. Lett.*, 20, 251, 1981.
108. **Nair, T. V. R., Grover, H. L., and Abrol, Y. P.**, Nitrogen metabolism of the upper three leaf blades of wheat at different soil nitrogen levels. II. Protease activity and mobilization of reduced nitrogen to the developing grains, *Physiol. Plant*, 42, 293, 1978.
109. **Drivdahl, R. H. and Thimann, K. V.**, Proteases of senescing oat leaves. I. Purification and general properties, *Plant Physiol.*, 59, 1059, 1977.
110. **Frith, G. J. T., Gordon, K. H. J., and Dalling, M. J.**, Proteolytic enzymes in green wheat leaves. I. Isolation on DEAE-cellulose of several proteinases with acid pH optima, *Plant Cell Physiol.*, 19, 491, 1978.
111. **Miller, B. L. and Huffaker, R. C.**, Partial purification and characterization of endoproteinases from senescing barley leaves, *Plant Physiol.*, 68, 930, 1981.

112. **Thomas, H.**, Leaf senescence in a non-yellowing mutant of *Festuca pratensis*. II. Proteolytic degradation of thylakoid and stroma polypeptides, *Planta,* 154, 219, 1982.
113. **Keist, M. and Feller, U.**, Changes of peptide hydrolase activities in wheat leaves during senescence, *Experientia,* 39, 649, 1983.
114. **Miller, B. L. and Huffaker, R. C.**, Hydrolysis of ribulose-1,5-bisphosphate carboxylase by endoproteinases from senescing barley leaves, *Plant Physiol.,* 69, 58, 1982.
115. **Peoples, M. B., Frith, G. J. T., and Dalling, M. J.**, Proteolytic enzymes in green wheat leaves. IV. Degradation of ribulose 1,5-bisphosphate carboxylase by acid proteinases isolated on DEAE-cellulose, *Plant Cell Physiol.,* 20, 253, 1979.
116. **Wells, J. R. E.**, Characterisation of three proteolytic enzymes from french beans, *Biochim. Biophys. Acta,* 167, 388, 1968.
117. **Peoples, M. B., Pate, J. S., and Atkins, C. A.**, Mobilization of nitrogen in fruiting plants of a cultivar of cowpea, *J. Exp. Bot.,* 34, 563, 1983.
118. **Van der Wilden, W., Segers, J. H. L., and Chrispeels, M. J.**, Cell walls of *Phaseolus vulgaris* leaves contain the azocoll-digesting proteinase, *Plant Physiol.,* 73, 576, 1983.
119. **Holzer, H. and Heinrich, P. C.**, Control of proteolysis, *Annu. Rev. Biochem.,* 49, 63, 1980.
120. **Hershko, A. and Ciechanover, A.**, Mechanisms of intracellular protein breakdown, *Annu. Rev. Biochem.,* 51, 335, 1982.
121. **Watanabe, T. and Kondo, N.**, The change in leaf protease and protease inhibitor activities after supplying various chemicals, *Biol. Plant.,* 25, 100, 1983.
122. **Cooke, R. J., Grego, S., Roberts, K., and Davies, D. D.**, The mechanism of deuterium oxide-induced protein degradation in *Lemna minor, Planta,* 148, 374, 1980.
123. **Matile, Ph.**, Protein degradation, in *Encyclopedia of Plant Physiology,* (New Series), Vol. 14A, Boulter, D. and Parthier, B., Eds., Springer-Verlag, Berlin, 1982, 169.
124. **Wittenbach, V. A., Lin, W., and Hebert, R. R.**, Vacuolar localization of proteases and degradation of chloroplasts in mesophyll protoplasts from senescing primary wheat leaves, *Plant Physiol.,* 69, 98, 1982.
125. **Streit, L. and Feller, U.**, Inactivation of N-assimilating enzymes and proteolytic activities in wheat leaf extracts: effect of pyridine nucleotides and of adenylates, *Experientia,* 38, 1176, 1982.
126. **Streit, L. and Feller, U.**, Nitrogen-metabolizing enzymes from bean leaves (*Phaseolus vulgaris* L.): stability *"in vitro"* and susceptibility to proteolysis, *Z. Pflanzenphysiol.,* 111, 19, 1983.
127. **Thomas, H.**, Leaf senescence in a non-yellowing mutant of *Festuca pratensis*. II. Proteolytic degradation of thylakoid and stroma polypeptides, *Planta,* 154, 219, 1982.
128. **Batt, R. G. and Wallace, W.**, A comparison of the effect of trypsin and a maize root proteinase on nitrate reductase and other enzymes from maize, *Biochim. Biophys. Acta,* 744, 205, 1983.
129. **Wallace, W.**, Effects of a nitrate reductase inactivating enzyme and NAD(P)H on the nitrate reductase from higher plants and *Neurospora, Biochim. Biophys. Acta,* 377, 239, 1975.
130. **Yamaya, T., Solomonson, L. P., and Oaks, A.**, Action of corn and rice-inactivating proteins on a purified nitrate reductase from *Chlorella vulgaris, Plant Physiol.,* 65, 146, 1980.
131. **Wray, J. L. and Kirk, D. W.**, Inhibition of NADH-nitrate reductase degradation in barley leaf extracts by leupeptin, *Plant Sci. Lett.,* 23, 207, 1981.
132. **Guiz, C., Hirel, B., Shedlofsky, G., and Gadal, P.**, Occurrence and influence of light on the relative proportions of two glutamine synthetases in rice leaves, *Plant Sci. Lett.,* 15, 271, 1979.
133. **Bond, J. S. and Offermann, M. K.**, Initial events in the degradation of soluble cellular enzymes: factors affecting the stability and proteolytic susceptibility of fructose-1,6-bisphosphate aldolase, *Arch. Biol. Med. Germ.,* 40, 1365, 1981.
134. **Barton, R.**, Fine structure of mesophyll cells in senescing leaves of *Phaseolus, Planta,* 71, 314, 1966.
135. **Dalling, M. J., Tang, A. B., and Huffaker, R. C.**, Evidence for the existence of peptide hydrolase activity associated with chloroplasts isolated from barley mesophyll protoplasts, *Z. Pflanzenphysiol.,* 111, 311, 1983.
136. **Thomas, H.**, Leaf senescence in a non-yellowing mutant of *Festuca pratensis*. I. Chloroplast membrane polypeptides, *Planta,* 154, 212, 1982.
137. **Thomas, H.**, Proteolysis in Senescing Leaves, Monograph 9, British Plant Growth Regulator Group, 1983, 45.
138. **Kaur-Sawhney, R., Shih, L., Cegielska, T., and Galston, A. W.**, Inhibition of protease activity by polyamines. Relevance for control of leaf senescence, *FEBS Lett.,* 145, 345, 1982.
139. **Kaur-Sawhney, R., Shih, L., Flores, H. E., and Galston, A. W.**, Relation of polyamine synthesis and titer to aging and senescence in oat leaves, *Plant Physiol.,* 69, 405, 1982.
140. **Burlini, N., Tortora, P., Hanozet, G. M., Vincenzini, M. T., Vanni, P., and Guerritore, A.**, Susceptibility to proteinases of yeast enzymes selectively modified by fatty acids, *Biochim. Biophys. Acta,* 708, 225, 1982.
141. **Berger, D., Vischer, T. L., and Micheli, A.**, Induction of proteolytic activity in serum by treatment with anionic detergents and organic solvents, *Experientia,* 39, 1109, 1983.

142. **Salgo, A. and Feller, U.**, Effect of low molecular weight compounds on proteolytic inactivation of glucose-6-phosphate dehydrogenase, *Plant Physiol.*, 72 (Suppl.), 35, 1983.
143. **Thomas, H.**, Control of chloroplast demolition during leaf senescence, in *Plant Growth Substances*, Wareing, P. F., Ed., Academic Press, London, 1982, 559.
144. **Maccioni, R. B. and Seeds, N. W.**, Limited proteolysis of tubulin: nucleotide stabilizes an active conformation, *Biochemistry*, 22, 1567, 1983.
145. **Waxman, L. and Goldberg, A. L.**, Protease La from *Escherichia coli* hydrolyzes ATP and proteins in a linked fashion, *Proc. Natl. Acad. Sci. U.S.A.*, 79, 4883, 1982.
146. **Desautels, M. and Goldberg, A. L.**, Demonstration of an ATP-dependent, vanadate-sensitive endoprotease in the matrix of rat liver mitochondria, *J. Biol. Chem.*, 257, 11673, 1982.
147. **Rapoport, S., Dubiel, W., and Müller, M.**, Characteristics of an ATP-dependent proteolytic system of rat liver mitochondria, *FEBS Lett.*, 147, 93, 1982.
148. **Hershko, A., Heller, H., Elias, S., and Ciechanover, A.**, Components of ubiquitin-protein ligase system. Resolution, affinity purification, and role in protein breakdown, *J. Biol. Chem.*, 258, 8206, 1983.
149. **Speiser, S. and Etlinger, J. D.**, ATP stimulates proteolysis in reticulocyte extracts by repressing an endogenous protease inhibitor, *Proc. Natl. Acad. Sci. U.S.A.*, 80, 3577, 1983.
150. **Hammond, J. B. W. and Preiss, J.**, ATP-dependent proteolytic activity from spinach leaves, *Plant Physiol.*, 73, 902, 1983.
151. **Liu, X.-Q. and Jagendorf, A. T.**, ATP-dependent proteolysis in pea chloroplasts, *FEBS Lett.*, 166, 248, 1984.
152. **Malek, L., Bogorad, L., Ayers, A. R., and Goldberg, A. L.**, Newly synthesized proteins are degraded by an ATP-stimulated proteolytic process in isolated pea chloroplasts, *FEBS Lett.*, 166, 253, 1984.
153. **Ciechanover, A., Finley, D., and Varshavsky, A.**, Ubiquitin dependence of selective protein degradation demonstrated in the mammalian cell cycle mutant ts85, *Cell*, 37, 57, 1984.
154. **Finley, D., Ciechanover, A., and Varshavsky, A.**, Thermolability of ubiquitin-activating enzyme from the mammalian cell cycle mutant ts85, *Cell*, 37, 43, 1984.
155. **Hershko, A., Leshinsky, E., Ganoth, D., and Heller, H.**, ATP-dependent degradation of ubiquitin-protein conjugates, *Proc. Natl. Acad. Sci. U.S.A.*, 81, 1619, 1984.
156. **Tanaka, K., Waxman, L., and Goldberg, A. L.**, Vanadate inhibits the ATP-dependent degradation of proteins in reticulocytes without affecting ubiquitin conjugation, *J. Biol. Chem.*, 259, 2803, 1984.
157. **Goldberg, A. L. and Boches, F. S.**, Oxidized proteins in erythrocytes are rapidly degraded by the adenosine triphosphate-dependent proteolytic system, *Science*, 215, 1107, 1982.
158. **Mattoo, A. K., Hoffman-Falk, H., Marder, J. B., and Edelman, M.**, Regulation of protein metabolism: coupling of photosynthetic electron transport to *in vivo* degradation of the rapidly metabolized 32-kilodalton protein of the chloroplast membranes, *Proc. Natl. Acad. Sci. U.S.A.*, 81, 1380, 1984.

Chapter 4

# ROLE OF PROTEOLYTIC ENZYMES IN THE POST-TRANSLATIONAL MODIFICATION OF PROTEINS

## J. Michael Lord and Colin Robinson

## TABLE OF CONTENTS

| | | |
|---|---|---|
| I. | Introduction | 70 |
| II. | Storage Protein Processing | 70 |
| III. | Lectin Processing | 73 |
| IV. | Post-Translational Processing of Cytoplasmically Synthesized Chloroplast Polypeptides | 74 |
| | A. Purification and Assay of the Processing Activity | 75 |
| | B. Physical Properties of the Processing Enzyme | 75 |
| | C. Two-Step Processing of P20 | 76 |
| | D. Specificity of the Processing Enzyme | 77 |
| References | | 77 |

## I. INTRODUCTION

Plant cells, like all eukaryotic cells, are characterized by a complex intracellular organization that defines several discrete compartments within the cell. The compartments or organelles are limited and defined by membranes and may be identified and differentiated from each other on the basis of their morphological and biochemical properties. The functional properties of the organelles are due in large part to the proteins that are uniquely housed within them. Eukaryotic cells have evolved a complex series of cotranslational and posttranslational events, which achieve the transfer of polypeptides from their sites of synthesis to their sites of function.[1] In the case of organellar proteins that are not encoded and synthesized within the organelles themselves, synthesis is either accompanied or rapidly followed by transport of the proteins across a membrane. Certain proteins such as mammalian secretory proteins or plant storage proteins are transported across the endoplasmic reticulum membrane during their synthesis on bound ribosomes. The messenger RNAs for such proteins usually encode a predominantly hydrophobic sequence of amino acids immediately to the 3' side of the initiation codon.[2] This N-terminal amino acid extension is termed a signal sequence, and its role, after interaction with a cytoplasmic signal recognition particle,[3] is to bind the ribosome to the cytoplasmic face of the endoplasmic reticulum.[4] The nascent proteins are then vectorially discharged across the membrane as the chain elongates (cotranslational transport). As the growing polypeptide extends into the lumen of the endoplasmic reticulum, the signal sequence is removed by a processing enzyme on the luminal surface of the membrane. This process can be reconstructed in a cell-free system when isolated rough microsomal vesicles translocate, process, glycosylate, and sequester nascent proteins.[5]

While secretory proteins and the proteins of organelles such as lysosomes and plant protein bodies are transported across a membrane during their synthesis, other organellar proteins are transported across a membrane after synthesis has been completed and they have been released from the ribosome.[1] Many of the proteins found in chloroplasts and mitochondria fall into this latter class.[6] These organelles contain their own genetic systems and are able to synthesize a small proportion of their constituent proteins. Most of the soluble and membrane proteins that become localized in these organelles are, however, initially synthesized on free cytoplasmic ribosomes and are transiently present in the cytoplasm. The proteins to be transported are usually, but not always, released into the cytoplasm as higher molecular weight precursors, which have an N-terminal amino acid extension. Import of the precursors into the organelles is accompanied by proteolytic processing that removes the N-terminal amino acid extension.[7] This processing step is achieved by soluble enzymes present in the mitochondrial matrix[8] or the chloroplast stroma.[9]

A second type of post-translational proteolytic activity is known to occur during the later stages of assembly of several plant storage proteins and lectins. Although these proteins are transported across the endoplasmic reticulum during synthesis, a step which normally involves cotranslational removal of the signal sequence,[1] they are still in precursor form during subsequent intracellular transport. Assembly of the native proteins in the protein bodies apparently requires a second, posttranslational processing step to release the constituent polypeptides of the mature proteins from the precursor.

## II. STORAGE PROTEIN PROCESSING

Plant seeds contain a high content of proteins known as storage proteins, which are rapidly degraded during the early stages of germination and seedling establishment (see Chapters 1 and 2). The storage proteins may account for up to 80% of the total protein in mature cells. About 70% of man's intake of proteins comes directly from seeds.[10] Not surprisingly, therefore, the occurrence, biosynthesis, and structure of seed storage proteins is currently attracting considerable attention. The most widely studied seeds are those of legumes, which store proteins in cotyledonary cells, and cereals, which store their reserves in the endosperm cells.

Legume cotyledons specialize in the production of two major storage globulins, legumin and vicilin.[11-13] Legumin is a high-molecular-weight (340,000 to 390,000; 11 to 12S) protein believed to consist of six distinct heterodimers. Each heterodimer contains a larger, acidic polypeptide chain ($M_r$ about 40,000) joined by disulfide bonds to a smaller basic polypeptide ($M_r$ about 20,000).[14,15] The dissociation of the holoprotein can be represented as follows:

$$\text{Holoprotein} \xrightarrow[\text{sulfate}]{\text{sodium dodecyl}} 6 \text{ DIMERS} \xrightarrow{\text{dithiothreitol}} 6 \text{ LARGE} + 6 \text{ SMALL POLYPEPTIDES}$$

Both the acidic and basic subunits are heterogeneous in charge and molecular weight.[16] This heterogenicity cannot be attributed to varying accounts of carbohydrate side chains since legumin polypeptides are not glycosylated.[17]

Vicilin is a storage protein of molecular weight 150,000 to 180,000 (7 to 9S).[12] In contrast to legumin, vicilin is N-glycosylated.[18] Vicilin from *Phaseolus vulgaris*[19] and *Glycine max*[20] contains three major subunits in the molecular weight range 40,000 to 60,000. The subunit composition of *Pisum sativum* vicilin is considerably more complex, and electrophoretic analysis has resolved ten major subunits with approximate molecular weights of 75,000, 70,000, 50,000 (a group of subunits), 34,000, 30,000, 25,000, 18,000, 14,000, 13,000, and 12,000.[21] The $M_r$ 50,000 group can be electrophoretically resolved into at least four components,[22] and an even more complex pattern emerges during isoelectric focusing.[23]

Recently a third storage protein, distinct from legumin and vicilin, has been isolated from pea seeds.[24] This protein, named convicilin, is a tetrameric globulin of molecular weight 290,000, with a subunit molecular weight of 71,000.

While all legumes contain both 11S legumin and 7S vicilin storage proteins, the respective proportions of these proteins can vary markedly between different plants. For example, the legumin/vicilin ratio is 4:1 in seeds of some *Vicia faba* varieties and only 1:9 in *Phaseolus*.[25]

Legumes storage proteins accumulate in cotyledonary cells in single membrane delimited protein bodies. Extensive studies have recently established that the globulin polypeptides are

1. Synthesized on membrane-bound polysomes[26-28]
2. Segregated by and transiently present in the rough endoplasmic reticulum[29-33]
3. Transported intracellularly, probably via the Golgi apparatus[34,35]
4. Finally located in protein bodies[35]

A major difference between legumin and vicilin type storage proteins of legumes is that the latter group are N-glycosylated, while the former are not. Thus pea vicilin, soybean β-conglycinin, and bean phaseolin contain 4 to 6% covalently attached carbohydrate consisting primarily of mannose (Man) and *N*-acetylglucosamine (GlcNAc).[36,37] The initial glycosylation reaction occurs cotranslationally[38,39] and involves the "en bloc" transfer of an oligosaccharide ($-[\text{GlcNAc}]_2[\text{Man}]_9[\text{Glc}]_3$) from a lipid carrier to the nascent polypeptide chain.[40,41] The glycosyltransferases responsible for this step are exclusively located in the endoplasmic reticulum,[42,43] and newly synthesized vicilin transiently present in the luminal space of the endoplasmic reticulum is already glycosylated.[31] In studies of vicilin synthesis in the presence of tunicamycin, the N-glycosylation is not necessary for synthesis or intracellular transport.[30,31] Further, nonglycosylated vicilin remains an effective substrate for the endoproteolytic enzymes in the protein bodies that generate the characteristic vicilin subunits[44] (see below).

Cereals usually differ from legumes in terms of their major storage proteins (Chapter 1). They are characterized by the production of prolamins (proteins rich in proline and glutamine; insoluble in water and salt solutions, but soluble in aqueous alcohol) and glutelins (insoluble in water, salt solutions, and aqueous alcohol; soluble in dilute acid or alkali.)[45,46] Oat is an exception to this general rule in having a globulin as its major storage protein.[47] Zein, the major storage protein of maize endosperm, consists of a group (up to 28) of similar polypeptides encoded by a multigene family.[48] Cereal storage proteins, like those of legumes, are translated on ribosomes attached to the endoplasmic reticulum. A signal peptide is cleaved cotranslationally and the nascent polypeptides are secreted into the lumen of the endoplasmic reticulum.[49-53]

A major feature of many, but by no means all, storage proteins is that they are encoded in the form of preproproteins. In such cases, generation of the characteristic storage protein subunits involves two readily distinguished proteolytic processing steps. The first of these occurs cotranslationally and results in the cleaveage of the N-terminal signal sequence.[54] The second step involves the endoproteolytic cleavage of the segregated proprotein. This posttranslational step is believed to occur in the protein bodies and takes place at least 1 hr and up to many hours after the protein has been synthesized.[30,31]

All proteins of the legumin type, including oat seed globulin, are initially synthesized as proproteins in which the acidic and basic subunit sequences are present in a single polypeptide chain. Endoproteolytic cleavage yields the separated subunits that are disulfide-linked in the heterodimers characteristic of the mature proteins.[51,53,55-59]

At present little is known about the post-translational endoproteolytic cleavage of the proprotein precursors. The enzymes responsible have not been identified, but it seems clear that they occur and function in the protein bodies.[31,32] Analysis of cDNA clones encoding storage protein precursors has shown that the processing step does not simply entail cleavage of a single peptide bond. The coding sequence of the acidic subunit precedes that of the basic subunit in the mRNA, and the two coding sequences are joined by a short linking sequence. Precursor processing involves the excision of this linking peptide. The excised peptide appears to be six amino acids long in the case of pea seen legumin[60] and four amino acids long in the case of soybean glycinin.[37] The carboxyterminal amino acid of the linker peptide is asparagine in each case. These and other examples (see below) suggest that asparagine plays a prominent role in processing sites.

All precursors to legumin-like proteins are cleaved to yield the acidic and basic subunits of the mature proteins. Vicilin-like proteins, on the other hand, require differing degrees of processing in different plant species. French bean phaseolin polypeptides, for example, appear to undergo little if any post-translational processing, the initial translation products being very similar in size to the mature protein subunits.[61] In contrast, pea vicilin consists of ten major subunits at least seven of which arise by post-translational endoproteolytic processing of primary translational products. Processing occurs in the protein bodies over a period 6 to 24 hr after translation.[31,32] The vicilin subunits of $M_r$ less than 50,000 show extensive sequence homology to regions of the 50,000 $M_r$ polypeptides from which they are now known to be derived.[62] These smaller vicilin subunits are formed by cleavage at one or both of two processing sites present in some, but not all, the 50,000 $M_r$ group of polypeptides.[62,63] A general scheme for the derivation of vicilin subunits from 5000 $M_r$ precursor polypeptides has been presented[62,63] and is illustrated here in Figure 1. Molecular weights given in Figure 1 are only approximate, and additional vicilin subunits will result when the derived fragments are glycosylated. Cleavage at both sites generates the smaller vicilin polypeptides, while cleavage at one site, but not at the other, generates intermediate sized polypeptides. It is not clear what factors determine if and where proteolytic cleavage

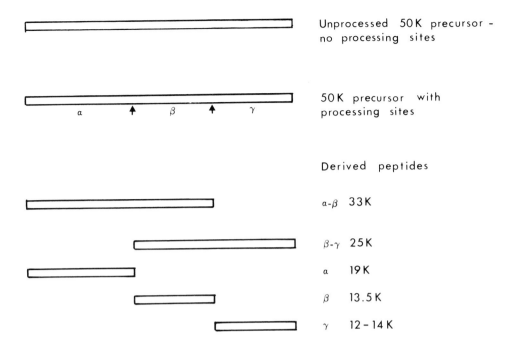

FIGURE 1. Postulated derivation of vicilin polypeptides from the $M_r$ 50,000 precursor.

occurs. However, since the 50,000 $M_r$ group of polypeptides represents a large proportion of mature vicilin, the extent to which proteolytic processing occurs is obviously limited. It seems likely that while some members of the 50,000 $M_r$ polypeptide group are completely processed at one or both sites, other members are not processed at all. The most likely factor in determining whether proteolysis occurs seems to be the amino acid sequence at the putative processing site.[64] The carboxy-terminal amino acid of the cleaved β-fragment is asparagine, and where the sequence Lys-Glu-Asn is present, cleavage at the β:γ site occurs on the carboxyl side of the asparagine residue. In other precursor molecules, where cleavage does not occur, this Lys-Glu-Asn sequence is replaced with Gly-Leu-Arg. Comparative data does not as yet exist for the α:β processing site. Vicilin cDNA clones have been sequenced and found to have Gly-Leu-Arg or Ser-Leu-Lys at the potential cleavage site, but it is believed that they represent sequences that are not processed at this site, particularly as Gly-Leu-Arg prevents cleavage when present at the β:γ cleavage site.[60]

In addition to endoproteolytic cleavage, post-translational processing of vicilin also involves the removal of a carboxy-terminal peptide. A comparison of the carboxy-terminus derived from a cDNA clone containing the complete coding sequence for preprovicilin with the amino acid sequence of mature vicilin revealed that a peptide 12 amino acids in length is removed from the carboxy-terminus.[60]

The studies described above have clearly shown that certain plant storage proteins undergo posttranslational processing within the protein bodies. A more detailed understanding of these processing events must await the identification and characterization of the proteolytic enzymes responsible.

## III. LECTIN PROCESSING

Plant seed lectins are carbohydrate-binding proteins of widespread occurrence, but unknown natural function.[65] These proteins are often abundantly present in seeds and are normally housed, together with the storage proteins, within the protein bodies. Lectins and storage proteins are simultaneously synthesized during seed development[66] and, not surprisingly, exhibit many common features.[67] Several lectins are heterodimers and the mechanism of their biosynthesis and intracellular transport is analogous to that of legumin-type proteins. These lectins are initially synthesized as preprolectin precursors that are cleaved both cotranslationally to remove a hydrophobic N-terminal leader sequence and post-translationally to yield the two different subunits of the mature lectins. The synthesis of favin from *V. faba* and pea seed lectin, both of which contain two distinct subunits, have recently been shown to fit this generalized scheme.[68,69] The larger (β) subunit of favin is N-glycosylated and the initial glycosylation event occurs cotranslationally during the synthesis of the preprolectin.[68] A study of pea lectin synthesis and intracellular transport has shown that the newly synthesized lectin precursor is transported from the endoplasmic reticulum to the protein bodies as the prolectin molecule. Post-translational processing of the pea lectin precursor takes place in the protein bodies.[67] The amino acid sequence deduced from the nucleotide sequence of cloned DNA complementary to pea lectin mRNA has shown that the endoproteolytic cleavage step that yields the two lectin subunits involves the excision of a six amino acid linking sequence that joins the two subunits in the proprotein.[69] Asparagine was identified as the carboxy-terminal amino acid of the excised peptide. Interestingly, the putative cleavage site that yields the two favin subunits is also on the carboxyl side of an asparagine residue.[70]

Pea lectin mRNA apparently encodes four amino acids at the carboxy-terminus of the subunit,[69] which are not present in the mature subunit.[71] These residues may also be removed post-translationally.

Endoproteolytic cleavage is not a feature of the synthesis of all legume lectins: concanavalin A (from *Canavalia ensiformis*) is a tetramer composed of a single subunit type.[72] Nor is endoproteolytic processing confined to lectins from legumes. The castor bean lectins, *Ricinus communis* agglutinin and the toxic lectin ricin, are also heterodimers that are initially synthesized as proprotein precursors.[73,74] In this case, post-translational cleavage of the precursor involves the excision of a 12 amino acid linking region which, once again, has asparagine at its carboxy-terminus.[75] A similar mechanism probably occurs for rice (*Oryza sativa*) lectin and will doubtless prove to be the case for other plant lectins.

At present nothing is known about the protein body endoprotease, which cleaves lectin precursors. Emerging evidence suggests that, as in the case of storage proteins, asparagine residues play an important role in defining at least one of the potential cleavage sites.

## IV. POST-TRANSLATIONAL PROCESSING OF CYTOPLASMICALLY SYNTHESIZED CHLOROPLAST POLYPEPTIDES

Chloroplast polypeptides originate from two distinct subcellular sources in the cells of higher plants and algae. A proportion of the polypeptides (about 20%) are encoded by chloroplast DNA and are synthesized within the organelle, while the remainder are imported after synthesis on cytoplasmic ribosomes.[76] The transport of cytoplasmically synthesized polypeptides into chloroplasts takes place post-translationally, i.e., after release of the polypeptide from the ribosome, and, therefore, differs fundamentally from the cotranslational insertion of secretory and protein body polypeptides into the endoplasmic reticulum.[77,78]

The synthesis in vitro of nuclear-encoded chloroplast polypeptides has been studied by translation of leaf cytoplasmic polyadenylated RNA in cell-free protein-synthesising systems, followed by immunoprecipitation of a given translation product. A number of imported polypeptides have been analyzed by this method, including the small subunit of ribulose bisphosphate carboxylase (RuBPCase),[77,79] the chlorophyll a/b binding protein,[80] plastocyanin,[81] ferredoxin-NADP oxidoreductase,[81] and the phosphate translocator of the inner envelope.[82] In each case, the polypeptide is initially synthesized as a high molecular weight

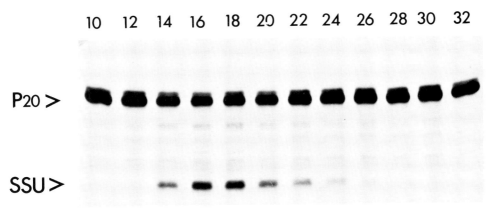

FIGURE 2. Elution of chloroplast processing activity from a DEAE-Sephacel column. A partially purified fraction from a stromal extract of pea chloroplasts was adsorbed to a column of DEAE-Sephacel, and eluted with a linear salt gradient. A number of eluate fractions were assayed for processing activity by incubation at 27°C with $^{35}$S-labeled small subunit precursor synthesized in a wheat-germ extract. Incubation was for 60 min, after which the samples were analysed by SDS-PAGE, followed by fluorography. Symbols: P20, small subunit precursor; SSU, mature small subunit. Other bands are endogenous wheat-germ products. Fraction numbers are indicated above the tracks.

precursor, the extension sequences ranging in size from 2,000 to 15,000 daltons. Although cell-free synthesis of only a small proportion of nuclear-encoded chloroplast polypeptides has been studied, it seems likely that most, if not all, of the remainder are also synthesized in precursor form.

The transport of these polypeptides into chloroplasts can be reconstituted in vitro by the addition of intact isolated chloroplasts to the cell-free translation products of leaf polyadenylated RNA. Precursors such as those listed above are taken up the chloroplasts, processed to the mature form, and segregated into the correct organellar compartment.[77,78] It is not known whether the processing event takes place during transport across the envelope membranes or shortly after import into the organelle, since no imported, unprocessed precursor molecules have been detected using this assay system.

The synthesis, transport, and processing of the precursor to the small subunit of RuBPCase have been most intensively studied. This precursor (P20, $M_r$ 20,000) is taken up by intact isolated chloroplasts and converted to the mature size ($M_r$ 14,000).[77] The protease responsible for this processing reaction has been partially purified from pea leaf chloroplasts,[83,84] and the characteristics of the enzyme are described below.

### A. Purification and Assay of the Processing Activity

A soluble stromal extract is prepared by lysis of washed chloroplasts, followed by centrifugation to remove envelope and thylakoid membranes. The processing activity is then purified by ammonium sulfate fractionation, gel filtration, and ion-exchange chromatography. The most highly purified fractions display 6 to 10 bands on a silver-stained SDS-polyacrylamide gel.

The enzymic activity is assayed by the ability to process ($^{35}$S)-labeled P20 to the mature size as judged by SDS-PAGE. The labeled precursor is prepared by cell-free translation of P20 mRNA, which has been purified by hybridization to cloned cDNA for small subunit. The elution of the processing activity from a column of DEAE-Sephacel is illustrated in Figure 2.

### B. Physical Properties of the Processing Enzyme

The partially purified enzyme has a molecular weight of approximately 180,000 as determined by calibrated gel filtration. It is active between pH 6.5 and 10.0, with an optimum

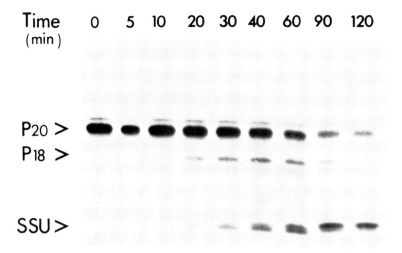

FIGURE 3. Time course of processing of small subunit precursor. Partially purified processing enzyme was incubated with $^{35}$S-labeled small subunit precursor at 27°C. At various time points, aliquots were removed, made 1% in SDS and boiled for 2 min. Samples were analyzed by SDS-PAGE followed by fluorography. Symbols: P20, small subunit precursor; P18, 18,000 $M_r$ processing intermediate; SSU, mature small subunit. Sample times are given above the tracks.

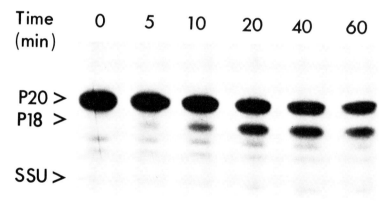

FIGURE 4. Time course of processing of iodoacetate-treated small subunit precursor. A time course of processing of small subunit precursor was carried out as described in the legend to Figure 3, except that the precursor was preincubated with 10 m$M$ iodoacetate for 30 min at 4°C before addition of the processing enzyme. Symbols as in Figure 3.

near 9.0 for the processing of P20. The enzyme is inhibited by metal-chelators such as EDTA and 1,10-phenanthroline, suggesting the presence of an essential metal ion in the active center. The serine protease inhibitor phenylmethylsulfonyl fluoride, and thiol reagents such as iodoacetate and $N$-ethylmaleimide, have no effect.

### C. Two-Step Processing of P20

The small subunit precursor from *Pisum* is processed to the mature size in two steps by the partially purified enzyme. During a time course of processing of P20, an intermediate form of molecular weight 18,000 appears and then declines as the precursor is processed to the mature size (Figure 3). The second cleavage is selectively inhibited if the precursor is preincubated with iodoacetate before addition of the processing enzyme (Figure 4). The effect of iodoacetate is believed to result from the involvement of cysteine in the second, but not in the first, scissile bond; the carboxymethylation of the cysteine residue at the second cleavage site probably renders the scissile bond inaccessible to the processing enzyme. The activities responsible for the two cleavage reactions copurify in a diverse range of purification procedures, strongly suggesting that a single enzyme carries out both reactions despite the presence of different residues at the cleavage sites. Iodoacetate-treated P20 is taken up by intact isolated chloroplasts and converted to the intermediate form, showing that the second cleavage reaction is not required for import into the organelle; the significance of the two-step processing mechanism is unclear.

### D. Specificity of the Processing Enzyme

The partially purified enzyme from *Pisum* processes the precursors of plastocyanin and ferredoxin-NADP oxidoreductase from *Triticum, Spinacia,* and *Hordeum* to the mature size, indicating that the enzyme is neither species- nor precursor-specific. However, it is not known whether a single enzyme is responsible for the processing of all imported precursors; a greater number of precursors must first be tested with the isolated processing enzyme, and the enzyme must be purified to homogeneity in order to exclude the possibility that more than one species of enzyme is present in the preparation.

The specific nature of the processing reaction carried out by the isolated enzyme is emphasized by two observations: the enzyme cleaves precursors of chloroplast proteins to the mature size but no further, even under conditions of enzyme excess, and no activity is displayed against any other, nonchloroplast proteins that have been tested to date. The basis for this reaction specificity is not known at present, though it is generally assumed that the extension sequence contains some, if not all, of the information specifying correct targeting of the processing enzyme. Whether the enzyme recognizes a consensus sequence or specific conformation in the precursors may become apparent when the extension sequences of a number of imported precursors have been deduced; to date only one extension sequence, that of the RuBPCase small subunit precursor, has been published.[85-87]

Although the precise mechanism of the processing reaction remains obscure, the contribution of a number of residues to the targeting site for the processing enzyme can be inferred from experiments in which certain residues are replaced by amino acid analogs. In these experiments, abnormal P20s were synthesized by translation of P20 mRNA in the presence of azetidine-2-carboxylic acid (a proline analog), canavanine (an arginine analog), or S-2-aminoethylcysteine (a lysine analog). Incorporation of any of these analogs into the P20 chain markedly inhibits processing by the isolated enzyme, suggesting that conformation (which proline residues influence considerably) and positively-charged amino acids are involved in targeting the processing enzyme.[88] However, these results do not indicate whether the essential residues are located in the extension sequence, the mature section of the sequence, or in both.

## REFERENCES

1. **Blobel, G.,** Intracellular protein topogenesis, *Proc. Natl. Acad. Sci. U.S.A.,* 77, 1496, 1980.
2. **Blobel, G. and Dobberstein, B.,** Transfer of proteins across membranes. II. Reconstitution of functional rough microsomes from heterologous components, *J. Cell Biol.,* 67, 852, 1975.
3. **Walter, P. and Blobel, G.,** Signal recognition particle contains a 7S RNA essential for protein translocation across the endoplasmic reticulum, *Nature (London),* 299, 691, 1982.
4. **Gilmore, R. and Blobel, G.,** Transient involvement of signal recognition particle and its receptor in the microsomal membrane prior to protein translocation, *Cell,* 35, 677, 1983.

5. **Lingappa, V. R., Lingappa, J. R., Prasad, R., Ebner, K. E., and Blobel, G.**, Coupled cell-free synthesis, segregation and core glycosylation of a secretory protein, *Proc. Natl. Acad. Sci. U.S.A.*, 75, 2338, 1978.
6. **Chua, N. H. and Schmidt, G. W.**, Transport of proteins into mitochondria and chloroplasts, *J. Cell Biol.*, 81, 461, 1979.
7. **Ellis, R. J. and Robinson, C.**, Post-translational transport and processing of cytoplasmically-synthesized precursors of organellar proteins, in *The Post-Translational Modification of Proteins*, Vol. 2, Freedman, R. B. and Hawker, H., Eds., Academic Press, New York, 1984, 25.
8. **Bohni, P. C., Baum, G., and Schatz, G.**, Import of proteins into mitochondria. Partial purification of a matrix-located protease involved in cleavage of mitochondrial precursor polypeptides, *J. Biol. Chem.*, 258, 4937, 1983.
9. **Smith, S. M. and Ellis, R. J.**, Processing of small subunit precursor of ribulose bisphosphate carboxylase and its assembly into whole enzyme are stromal events, *Nature (London)*, 278, 662, 1979.
10. **Spencer, D. and Higgins, T. J. V.**, Molecular aspects of seed protein biosynthesis, *Curr. Adv. Plant. Sci.*, 34, 1, 1979.
11. **Millard, A.**, Biochemistry of legume seed proteins, *Annu. Rev. Plant. Physiol.*, 26, 53, 1975.
12. **Derbyshire, E., Wright, D. J., and Boulter, D.**, Legumin and vicilin, storage proteins of legume seeds, *Phytochemistry*, 15, 3, 1976.
13. **Boulter, D.**, Proteins of legumes, in *Advances in Legume Systematics*, (Part 2), Polhill, R. M. and Raven, P. H., Eds., Royal Botanic Gardens, Kew, 1981, 501.
14. **Croy, R. R. D., Derbyshire, E., Krishna, T. G., and Boulter, D.**, Legumin of *Pisum sativum* and *Vicia faba*, *New Phytol.*, 83, 29, 1979.
15. **Krishna, T. G., Croy, R. R. D., and Boulter, D.**, Heterogeneity in subunit composition of the legumin of *Pisum sativum*, *Phytochemistry*, 18, 1879, 1979.
16. **Gatehouse, J. A., Croy, R. R. D., and Boulter, D.**, Isoelectric focusing properties and carbohydrate content of pea (*Pisum sativum*) legumin, *Biochem. J.*, 185, 497, 1980.
17. **Casey, R.**, Immunoaffinity chromatography as a means of purifying legumin from *Pisum* (pea) seeds, *Biochem. J.*, 177, 509, 1979.
18. **Basha, S. S. M. and Beevers, L.**, Glycoprotein metabolism in the cotyledons of *Pisum sativum* during development and germination, *Plant Physiol.*, 57, 93, 1976.
19. **Sun, S. M., McLeester, R. C., Bliss, F. A., and Hall, T. C.**, Reversible and irreversible dissociation of globulins from *Phaseolus vulgaris* seeds, *J. Biol. Chem.*, 249, 2118, 1974.
20. **Thanh, V. H. and Shibasaki, K.**, Major proteins in soybean seeds. Subunit structure of β-conglycinin, *J. Agric. Food Chem.*, 26, 692, 1978.
21. **Badenoch-Jones, J., Spencer, D., Higgins, T. J. V., and Millard, A.**, The role of glycosylation in storage protein synthesis in developing pea seeds, *Planta*, 153, 201, 1981.
22. **Thompson, J. A., Schroeder, H. E., and Tassie, A. M.**, Cotyledonary storage proteins in *Pisum sativum*. V. Further studies on molecular heterogeneity in the vicilin series of holoproteins, *Aust. J. Plant Physiol.*, 7, 271, 1980.
23. **Gatehouse, J. A., Croy, R. R. D., Morton, H., Tyler, M., and Boulter, D.**, Characterization and subunit structures of the vicilin storage proteins of pea *(Pisum sativum)*, *Eur. J. Biochem.*, 118, 627, 1981.
24. **Croy, R. R. D., Gatehouse, J. A., Tyler, M., and Boulter, D.**, The purification and characterization of a third storage protein (convicilin) from the seeds of pea (*Pisum sativum* L.), *Biochem. J.*, 191, 509, 1980.
25. **Carasco, J. F., Croy, R. R. D., Derbyshire, E., and Boulter, D.**, The isolation and characterization of the major polypeptides of the seed globulin of cowpea and their sequential synthesis in developing seeds, *J. Exp. Bot.*, 29, 309, 1978.
26. **Punchel, M., Muntz, K., Parthier, B., Aurich, O., Bassuner, R., Mantenfell, R., and Schmidt, P.**, RNA metabolism and membrane-bound polysomes in relation to globulin biosynthesis in cotyledons of developing field beans (*Vicia faba* L.), *Eur. J. Biochem.*, 96, 321, 1979.
27. **Bollini, R. and Chrispeels, M. J.**, The rough endoplasmic reticulum is the site of reserve protein synthesis in developing *Phaseolus vulgaris* cotyledons, *Planta*, 146, 487, 1979.
28. **Higgins, T. J. V. and Spencer, D.**, Precursor forms of pea vicilin subunits. Modification by microsomal membranes during cell-free translation, *Plant Physiol.*, 67, 205, 1981.
29. **Baumgartner, B., Tokuyasu, K. T., and Chrispeels, M. J.**, Immunocytochemical localization of phaseolin in the endoplasmic reticulum of developing bean (*Phaseolus vulgaris*) cotyledons, *Planta*, 150, 419, 1980.
30. **Bollini, R., Van der Wilder, W., and Chrispeels, M. J.**, A precursor of the reserve protein phaseolin is transiently associated with the endoplasmic reticulum of developing *Phaseolus vulgaris* cotyledons, *Physiol. Plant.*, 55, 82, 1982.

31. **Chrispeels, M. J., Higgins, T. J. V., Craig, S., and Spencer, D.**, Role of the endoplasmic reticulum in the synthesis of reserve proteins and the kinetics of their transport to protein bodies in developing pea cotyledons, *J. Cell Biol.*, 93, 5, 1982.
32. **Chrispeels, M. J., Higgins, T. J. V., and Spencer, D.**, Assembly of storage protein oligomers in the endoplasmic reticulum and processing of the polypeptides in the protein bodies of developing pea cotyledons, *J. Cell Biol.*, 93, 306, 1982.
33. **Hurkman, W. J. and Beevers, L.**, Sequestration of pea reserve proteins by rough microsomes, *Plant Physiol.*, 69, 1414, 1982.
34. **Chrispeels, M. J.**, Incorporation of fucose into the carbohydrate moiety of phytohaemagglutinin in developing *Phaseolus vulgaris* cotyledons, *Planta*, 157, 454, 1983.
35. **Chrispeels, M. J.**, The Golgi apparatus mediates the transport of phytohaemagglutinin to the protein bodies in bean cotyledons, *Planta*, 158, 140, 1983.
36. **Spencer, D.**, The physiological role of storage proteins in seeds, *Phil. Trans. R. Soc. London Ser. B*, 304, 275, 1984.
37. **Nielsen, N. C.**, The chemistry of legume storage proteins, *Phil. Trans. R. Soc. London Ser. B*, 304, 287, 1984.
38. **Kiely, M. L., McKnight, G. S., and Schimke, R. T.**, Studies on the attachment of carbohydrate to ovalbumin nascent chains in hen oviduct, *J. Biol. Chem.*, 251, 5490, 1976.
39. **Glabe, G. C., Hanover, J. A., and Lennarz, W. J.**, Glycosylation of ovalbumin nascent chains. The spatial relationship between translation and glycosylation, *J. Biol. Chem.*, 255, 9236, 1980.
40. **Hubbard, S. C. and Ivatt, R. J.**, Synthesis and processing of asparagine-linked oligosaccharides, *Annu. Rev. Biochem.*, 50, 555, 1981.
41. **Elbein, A. D.**, The role of lipid-linked saccharides in the biosynthesis of complex carbohydrates, *Encycl. Plant Physiol.*, 13B, 166, 1981.
42. **Lehle, L., Bowles, D. J., and Tanner, W.**, Subcellular site of mannosyl transfer to dolicholpyrophosphate in *Phaseolus aureus*, *Plant Sci. Lett.*, 11, 27, 1978.
43. **Nagahashi, J. and Beevers, L.**, Subcellular localization of glycosyltransferases involved in glycoprotein biosynthesis in the cotyledons of *Pisum sativum*, *Plant Physiol.*, 61, 451, 1978.
44. **Davey, R. A., Higgins, T. J. V., and Spencer, D.**, Homologies between two small subunits of vicilin from *Pisum sativum*, *Biochem. Int.*, 3, 595, 1981.
45. **Payne, P. I. and Rhodes, A. P.**, Cereal storage proteins: structure and role in agriculture and food technology, *Encycl. Plant Physiol.*, 14A, 346, 1982.
46. **Miflin, B. J., Field, J. M., and Shewry, P. R.**, Cereal storage proteins and their effects on technological properties, in *Seed proteins*, Danssant, J., Mosse, J., and Vaughan, J., Eds., Academic Press, London, 1983, 255.
47. **Petersen, D. M. and Smith, D.**, Changes in nitrogen and carbohydrate fractions in developing oat groats, *Crop Sci.*, 16, 67, 1976.
48. **Soave, C. and Salamini, F.**, Organization and regulation of Zein genes in maize endosperm, *Phil. Trans. R. Soc. London Ser. B.*, 304, 341, 1984.
49. **Larkins, B. A. and Hurkman, W. J.**, Synthesis and deposition of Zein in protein bodies of maize endosperm, *Plant Physiol.*, 62, 256, 1978.
50. **Burr, F. A. and Burr, B.**, *In vitro* uptake and processing of prezein and other maize preproteins by maize membranes, *J. Cell Biol.*, 90, 427, 1981.
51. **Brinegar, A. C. and Petersen, D. M.**, Synthesis of oat globulin precursors. Analogy to legume 11S storage protein synthesis, *Plant Physiol.*, 70, 1767, 1982.
52. **Yamagata, H., Sugimoto, T., Tanaka, K. and Kasai, Z.**, Biosynthesis of storage proteins in developing rice seedlings, *Plant Physiol.*, 70, 1094, 1982.
53. **Walburg, E. and Larkins, B. A.**, Oak seed globulin. Subunit characterization and demonstration of its synthesis as a precursor, *Plant Physiol.*, 72, 161, 1983.
54. **Bassinner, R., Wobus, U., and Rapport, T. A.**, Signal recognition particle triggers the translocation of storage globulin polypeptides from field beans (*Vicia faba* L.) across mammalian endoplasmic reticulum membrane, *FEBS Lett.*, 166, 314, 1984.
55. **Croy, R. R. D., Gatehouse, J. A., Evans, D. M., and Boulter, D.**, Characterization of the storage protein subunits synthesized *in vitro* by polyribosomes and RNA from developing pea (*Pisum sativum* L.). I. Legumin, *Planta*, 148, 44, 1980.
56. **Spencer, D. and Higgins, T. J. V.**, The biosynthesis of legumin in developing pea seeds, *Biochem. Int.*, 1, 502, 1980.
57. **Matlashewski, G. J., Adeli, K., Altosaar, I., Shewry, P. R., and Miflin, B. J.**, *In vitro* synthesis of oat globulin, *FEBS Lett.*, 145, 208, 1981.
58. **Sengupta, C., Deluca, V., Bailey, D. S., and Verma, D. P. S.**, Post-translational processing of 7S and 11S components of soybean storage proteins, *Plant Mol. Biol.*, 1, 19, 1981.
59. **Tumer, N. E., Thanh, V. H., and Nielsen, N. C.**, Purification and characterization of mRNA from soybean seeds. Identification of glycinin and β-conglycinin precursors, *J. Biol. Chem.*, 256, 8756, 1981.

60. **Boulter, D.,** Cloning of pea storage protein genes, *Phil. Trans. R. Soc. London Ser. B.*, 304, 323, 1984.
61. **Slighton, J. L., Sun, S. M., and Hall, T. C.,** Complete nucleotide sequence of a French bean storage protein gene: phaseolin, *Proc. Natl. Acad. Sci. U.S.A.*, 80, 1897, 1983.
62. **Spencer, D., Chandler, P. M., Higgins, T. J. V., Inglis, A., and Rubira, M.,** Sequence inter-relationships of the subunits of vicilin from pea seeds, *Plant Mol. Biol.*, 2, 259, 1983.
63. **Gatehouse, J. A., Lycett, G. W., Delanney, A. J., Croy, R. R. D., and Boulter, D.,** Sequence specificity of the post-translational proteolytic cleavage of vicilin, a seed storage protein of pea, *Biochem. J.*, 212, 427, 1983.
64. **Lycett, G. W., Delanney, A. J., Gatehouse, J. A., Gilroy, J., Croy, R. R. D., and Boulter, D.,** The vicilin gene family of pea (*Pisum sativum*): a complete cDNA coding sequence for preprovicilin, *Nucl. Acid Res.*, 11, 2367, 1983.
65. **Barondes, S. H.,** Lectins: their multiple endogenous cellular functions, *Annu. Rev. Biochem.*, 50, 207, 1981.
66. **Roberts, L. M. and Lord, J. M.,** Protein biosynthetic capacity in the endosperm tissue of ripening castor bean seeds, *Planta*, 152, 420, 1981.
67. **Higgins, T. J. V., Chrispeels, M. J., Chandler, P. M., and Spencer, D.,** Intracellular sites of synthesis and processing of lectin in developing pea cotyledons, *J. Biol. Chem.*, 258, 9550, 1983.
68. **Hemperley, J. J., Mostor, K. E., and Cunningham, B. A.,** *In vitro* translation and processing of a precursor form of favin, a lectin from *Vicia faba*, *J. Biol. Chem.*, 257, 7903, 1982.
69. **Higgins, T. J. V., Chandler, P. M., Zurawski, G., Button, S. C., and Spencer, D.,** The biosynthesis and primary structure of pea seed lectin, *J. Biol. Chem.*, 258, 9544, 1983.
70. **Hemperley, J. J. and Cunningham, B. A.,** Circular permutation of amino acid sequences among legume lectins, *Trends Biochem. Sci.*, 8, 100, 1983.
71. **Richardson, C., Behnke, W. D., Freisheim, J. H., and Blumenthal, K. M.,** The complete amino acid sequence of the α-subunit of pea lectin, *Biochem. Biophys. Acta*, 537, 310, 1978.
72. **Wang, J. L., Cunningham, B. A., and Edelman, G. M.,** Unusual fragments in the subunit structure of concanavalin A, *Proc. Natl. Acad. Sci. U.S.A.*, 68, 1130, 1971.
73. **Roberts, L. M. and Lord, J. M.,** Synthesis of *Ricinus communis* agglutinin. Co- and post-translational modification of agglutinin polypeptides, *Eur. J. Biochem.*, 119, 31, 1981.
74. **Butterworth, A. G. and Lord, J. M.,** Ricin and *Ricinus communis* agglutinin subunits are all derived from a single size polypeptide precursor, *Eur. J. Biochem.*, 137, 57, 1983.
75. **Lamb, F. I., Roberts, L. M., and Lord, J. M.,** unpublished data.
76. **Ellis, R. J.,** Chloroplast proteins: synthesis, transport and assembly, *Annu. Rev. Plant. Physiol.*, 32, 111, 1981.
77. **Highfield, P. E. and Ellis, R. J.,** Synthesis and transport of the small subunit of chloroplast ribulose bisphosphate carboxylase, *Nature (London)*, 271, 420, 1978.
78. **Chua, N-H. and Schmidt, G. W.,** Post-translational transport into intact chloroplasts of a precursor to the small subunit of ribulose-1,5-bisphosphate carboxylase, *Proc. Natl. Acad. Sci. U.S.A.*, 75, 6110, 1978.
79. **Dobberstein, B., Blobel, G., and Chua, N-H.,** *In vitro* synthesis and processing of a putative precursor for the small subunit of ribulose-1,5-bisphosphate carboxylase of *Chlamydomonas reinhardtii*, *Proc. Natl. Acad. Sci. U.S.A.*, 74, 1082, 1977.
80. **Apel, K. and Kloppstech, K.,** Light-induced appearance of mRNA coding for the apoprotein of the light-harvesting chlorophyll a/b protein, *Eur. J. Biochem.*, 85, 581, 1978.
81. **Grossman, A. R., Bartlett, S. G., Schmidt, G. W., Mullett, J. E., and Chua, N-H.,** Optimal conditions for post-translational uptake of proteins by isolated chloroplasts, *J. Biol. Chem.*, 257, 1558, 1982.
82. **Flugge, U. I. and Wessel, D.,** Cell-free synthesis of putative precursors for envelope membrane polypeptides of spinach chloroplasts, *FEBS Lett.*, 168, 255, 1984.
83. **Robinson, C. and Ellis, R. J.,** Transport of proteins into chloroplasts. Partial purification of a chloroplast protease involved in the processing of imported precursor polypeptides, *Eur. J. Biochem.*, 1984, in press.
84. **Robinson, C. and Ellis, R. J.,** Transport of proteins into chloroplasts. The precursor of ribulose bisphosphate carboxylase small subunit is processed to the mature size in two steps, *Eur. J. Biochem.*, 142, 343, 1984.
85. **Cashmore, A. R.,** Nuclear genes encoding the small subunit of ribulose-1,5-bisphosphate carboxylase, in *Genetic Engineering of Plants — an Agricultural Perspective*, Kosuge, T., Meredith, C. P., and Hollaender, A., Eds., Plenum Press, New York, 1983.
86. **Coruzzi, G., Broglie, R., Cashmore, A. R., and Chua, N-H.,** DNA sequence of cDNA clones encoding two chloroplast proteins, *J. Biol. Chem.*, 258, 1399, 1983.
87. **Berry-Lowe, S., McKnight, T. D., Shah, D. M., and Meagher, R. B.,** The nucleotide sequence, expression and evolution of one member of a multigene family encoding the small subunit of ribulose-1,5-bisphosphate carboxylase in soybean, *J. Mol. Appl. Genet.*, 1, 483, 1982.
88. **Robinson, C. and Ellis, R. J.,** unpublished data, 1984.

Chapter 5

# ROLE OF PROTEINASES IN THE REGULATION OF NITRATE REDUCTASE

### W. Wallace and A. Oaks

## TABLE OF CONTENTS

| | | |
|---|---|---|
| I. | Introduction | 82 |
| II. | Characteristics of Nitrate Reductase | 82 |
| | A. Structure of Nitrate Reductase | 82 |
| | B. Evidence for the High Rate of Turnover of Nitrate Reductase | 82 |
| III. | Macromolecules and Other Metabolites that Influence the Activity of Nitrate Reductase | 83 |
| | A. Inhibitors, Inactivators, and Proteinases | 83 |
| | B. Protector Molecules | 84 |
| | C. Reversible Inactivation by NADH | 84 |
| IV. | Degradation of Nitrate Reductase by Plant Proteinases | 85 |
| | A. Maize Root Proteinase | 85 |
| | B. Barley Leaf Proteinase | 86 |
| V. | Evaluation of Proteinases in the Regulation of Nitrate Reductase | 87 |
| References | | 87 |

## I. INTRODUCTION

Nitrate reductase (NR) is one of the most extensively studied plant enzymes,[1] yet we have only a relatively poor understanding of its structure and regulation. In recent studies, e.g., with the barley leaf,[2-4] where proteolysis has been minimized by selection of young leaf material, use of a high pH for the enzyme extraction, and inclusion of exogenous protein and a range of proteinase inhibitors in the enzyme isolation medium, good progress has been made in the preparation of small quantities of homogenous NR.[4,5] In this review, we present a structural model for higher plant NR and describe how it is attacked by proteinases. The role of such proteinases in the degradation of NR in the cell and in its overall regulation will also be evaluated.

## II. CHARACTERISTICS OF NITRATE REDUCTASE

### A. Structure of Nitrate Reductase

There now appears to be a general consensus[2,6-8] that the NADH-dependent NR (EC 1.6.6.1) in higher plants ($M_r$ approximately 200,000) consists (Figure 1) of two hemoflavoprotein subunits ($M_r$ about 100,000) and two molybdenum cofactors ($M_r < 1000$).[9] The molybdenum cofactor is noncovalently bound, has a reduced pterin ring in its structure, and is a constituent of several molybdoenzymes.[10]

*Neurospora* NR (NADPH-dependent, EC 1.6.6.3) appears to have a similar structure to the higher plant enzyme,[9] but in recent studies where proteolysis was minimized during its isolation,[11] the molecular weight of the monomer was found not to be 112,000 as reported earlier,[9] but 145,000. The *Chlorella* NR (EC 1.6.6.1) is a homotetramer of hemoflavoprotein subunits ($M_r$ 90,000), but dissociated at low enzyme concentration to a dimeric form.[12] Two species of NR, with molecular weights of about 166,000 and 433,000[8] have been isolated from maize leaves and were shown to be interconvertible.[8] In a recent study on spinach leaf NR (purified about 3000-fold) it was found to consist of two identical subunits, $M_r$ 130,000.[13] Howard and Solomonson[12] concluded that the variation reported in the physical parameters of NR from different sources may be due to factors such as enzyme concentration or ionic strength. Limited proteolysis is also likely to cause interference.[11]

The activity of NR can be measured by its overall reaction (NADH dependent reduction of $NO_3^-$) or by partial reactions of the enzyme complex (Figure 1). Treatment with mild heat or *p*-chloromercuribenzoate[14] results in loss of NADH-dependent cytochrome *c* reductase (CR) activity without effect on the reduced flavin (e.g., $FADH_2$-NR) or reduced viologen-dependent NR activity (e.g., BVH-NR). Alternatively, when $NO_3^-$ reduction is blocked, e.g., cyanide binding to molybdenum (Section III.C), NADH-CR activity can still be measured. In addition to CR associated with the NR complex, smaller CR species (also $NO_3^-$-inducible) are detected in plant extracts. Their relationship to NR will be discussed in Section IV.B.

### B. Evidence for the High Rate of Turnover of Nitrate Reductase

The main species of NR in higher plants is substrate inducible; increase in the amount of the enzyme resulting mainly from enhanced synthesis.[15] Recently it has also been established that the synthesis of the molybdenum cofactor is induced by nitrate.[16] NR cross-reacting material was not detected in root or shoot extracts from barley seedlings grown without nitrate,[4] but similar studies indicated an inactive form of NR in ammonium-grown *Chlorella* cells.[17]

In vivo decay values (half-life) for NR of $1\frac{1}{2}$ to 6 hr have been measured in several plants.[18-20] In these studies, synthesis of NR was blocked by cycloheximide[18,19] or tungstate.[20] With nitrate there was a small increase in the stability of the corn leaf and root enzyme.[20]

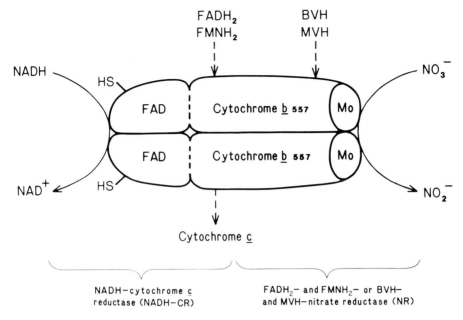

FIGURE 1. Structure and catalytic activities of nitrate reductase in higher plants.

Table 1
MACROMOLECULAR FACTORS WHICH CAUSE LOSS OF NITRATE REDUCTASE ACTIVITY

| Plant source | Molecular weight | Mechanism of action | Main characteristics | Ref. |
|---|---|---|---|---|
| Rice cell | 150 to 200,000 | Inhibition | Binds to NR-reversed by NADH | 23 |
| Soybean leaf | 31,000 | Inhibition | Light inactivated/dark activated | 24 |
| Maize root | 55 to 75,000 | Proteolysis | Serine proteinase | 22,25,40 |
| Barley leaf | 74,000 | Proteolysis | Thiol proteinase | 56 |
| Wheat leaf | 37,500 | Not known | Metal dependent, highest activity in dark | 29 |

Such relatively high decay rates have been confirmed with labeling techniques.[15,21] In tobacco cells, for example, it was demonstrated that NR had a relatively constant rate of degradation whether it was being induced, in a steady state, or showing net degradation.[15] During the synthesis and decay of NR in the corn plant, the partial activities of the NR complex varied in parallel.[20] Decrease in barley shoot NR protein (cross-reacting material) was slightly slower than loss of NR activity when nitrate was removed. However, no immunologically reactive degradation products were detected.[4]

## III. MACROMOLECULES AND OTHER METABOLITES THAT INFLUENCE THE ACTIVITY OF NITRATE REDUCTASE

### A. Inhibitors, Inactivators, and Proteinases

In the attempt to stabilize higher plant NR, several macromolecular factors have been described that catalyze the loss of NR activity in vitro (Table 1). These are heat labile and appear to be proteins. A maize root component, first described as a NR-inactivating enzyme,[22] has now been shown to be a proteinase (Section IV.A). By contrast, inhibitors from the rice cell[23] and soybean leaf[24] mediate a reversible loss of NR activity. The rice inhibitor ($M_r$ 150 to 200,000)[23,25] causes a loss of NADH-CR, but not MVH-NR activity[26] (Table 2).

## Table 2
### ACTION OF INHIBITORS, PROTEINASES, AND INACTIVATORS ON *CHLORELLA* AND HIGHER PLANT NR

| | Source | | | Inactivation (%) of NR and its component activities[a] | | |
|---|---|---|---|---|---|---|
| Inhibitory factor | NR | Time (hr) | NADH-NR | $FADH_2$ or $FMNH_2$-NR | BVH or MVH-NR | Ref. |
| Rice cell inhibitor | Rice cell | 1 | 44 | 21 | 0 | 23 |
| Maize root proteinase | Maize scutellum | 1 | 79 | 47 | 0 | 42 |
| | | 3 | 100 | 70 | 40 | |
| Maize root proteinase | *Chlorella* | 3 | 100 | — | 0 | 26 |
| Trypsin | Maize scutellum | 1 | 45 | 35 | 30 | 42 |
| Trypsin | *Chlorella* | 3 | 60 | — | 0 | 26 |
| Barley leaf proteinase | Barley leaf | — | 70 | 55 | 53 | 56 |
| Wheat leaf inactivator[b] | Wheat leaf | 1 | 81 | 62 | 53 | 29 |

[a] See Figure 1.
[b] For definition of inactivator see Section III.A.

NADH protected rice cell NR against inactivation by the inhibitor and facilitated reactivation of NR.[27] In a separate study on rice roots, two NR inhibitors were identified, one of $M_r$ 46,000 and a $M_r$ 240,000 species, which was shown to be a complex with NR.[28] The soybean leaf inhibitor ($M_r$ 31,000) had no effect on NADH-CR activity but $FADH_2$-NR was inactivated.[24] This inhibitor, which was light inactivated, regained its activity in the dark and could have a role in diurnal fluctuations in the activity of NR.

In contrast to the irreversible inactivation of NR by proteinases and the reversible modulation of its activity by inhibitors, other inhibitory components have been characterized that gave an irreversible loss of NR activity, but had no detectable proteinase or peptidase activity, e.g., wheat leaf (Table 1).[29] To reduce the confusion in the literature, we recommend the term inactivator for such macromolecules.

### B. Protector Molecules

Macromolecules have also been described that stabilize NR. In the wheat leaf, two species were identified,[30] which counteract the action of the NR-inactivator (Section III.A). The cotton seed and young cotyledon have a NR-stabilizing molecule ($M_r > 12,000$),[31] which was considered to protect NR from physical denaturation in vitro. Casein and bovine serum albumin have also been used with great success to stabilize NR in plant extracts.[32] These proteins could be serving as an alternative substrate for endogenous proteinases, binding inhibitory molecules such as polyphenols and/or stabilizing the NR molecule.

A NR-activator molecule has also been identified in the maize scutellum.[33] It was trypsin insensitive and gave some activation of corn leaf or root NR, as well as protecting these enzymes against the maize root proteinase and rice cell inhibitor.[33]

### C. Reversible Inactivation by NADH

It has been demonstrated for several plant tissues that NR, when reduced with NADH (in the absence of $NO_3^-$), is converted to an inactive state.[34] It appears, as has been demonstrated for *Chlorella*,[35] that this is due to cyanide binding to the molybdenum component of the reduced enzyme. Reactivation of NR occurs after its oxidation with ferricyanide,[36] or exposure to blue light plus FAD.[37] Evidence has been obtained for the occurrence in vivo of NR in a partially inactive state that could be activated by these oxidation procedures.[34] In contrast to this inactivation of NR by NADH, there is also some evidence that under some circumstances NADH can stabilize NR in vitro.[38,39]

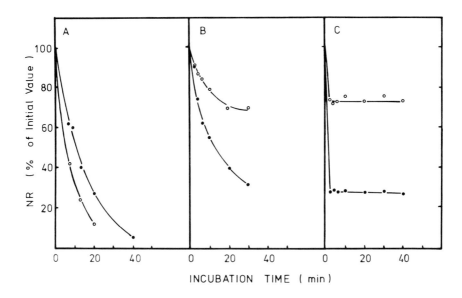

FIGURE 2. Action of trypsin (A), maize root proteinase (B), and flax cotyledon proteinase (C) on maize leaf nitrate reductase. NADH-NR (——●——) and MVH-NR (——○——).

## IV. DEGRADATION OF NITRATE REDUCTASE BY PLANT PROTEINASES

### A. Maize Root Proteinase

This proteinase, first characterized by its inactivation of NR and degradation of azocasein,[22] was subsequently purified independently in two laboratories.[25,40] Similar procedures were used except that Shannon and Wallace[40] used a DEAE-Sephadex® step, while Yamaya et al.[25] employed a final polyacrylamide gel electrophoresis step to obtain a homogeneous sample. There was some discrepancy in the molecular weight estimates, 54,000[40] vs. 75,000[25] (66,000 on SDS-PAGE), while in a separate study[41] a $M_r$ of 56,000 was obtained. The proteinase was inhibited by phenylmethylsulfonyl fluoride (in the presence of dithiotreitol) and appears to be a serine-proteinase.[40] It was active on alanyl-ester substrates, suggesting that it has the "tight" elastase type of active site.[42] Of two NR-inactivating proteins characterized in wheat leaves, one was also considered to be a serine-proteinase.[43]

The maize proteinase causes a preferential loss of the NADH-CR activity of NR, loss of NADH-NR activity occurring at a faster rate than that of $FADH_2$- or BVH-NR[36,42] (Table 2). The BVH-NR moiety was especially resistant to the proteinase; with maize scutella NR a lag of $1^1/_2$ hr occurred before any loss of BVH-NR activity.[42] Incubation of the scutellum NR with trypsin resulted in the same rate of loss of $FADH_2$- and BVH-NR as NADH-NR activity[42] (Table 2). A similar phenomenon was found with corn leaf NR[44] (Figure 2). When this enzyme was incubated with trypsin, there was the same rate of loss of MVH- and NADH-NR activity, but with the maize root proteinase less effect was observed on MVH-NR than on the overall NR complex. A flax proteinase[45] that degrades isocitrate lyase caused a rapid loss, 70 and 30%, respectively, of the NADH- and BVH-NR activity of the maize leaf enzyme, but no further inactivation with continued incubation. With *Chlorella* NR[26] neither trypsin nor the maize root proteinase caused any loss of MVH-NR activity, and with the latter there was some increase in activity as NADH-NR activity was lost.

In the absence of an adequate supply of a homogeneous sample of higher plant NR, structural changes in the enzyme resulting from proteolysis were investigated with *Chlorella*

NR.[26,46] After incubation with the maize root proteinase, *Chlorella* NR ($M_r$ 376,000; subunit $M_r$ 30,000). The tetrameric derivative retained MVH-NR activity and contained heme, but the smaller fragment, although it had FAD associated with it, had no activity. It appears that limited proteolysis results in the release of a small fragment ($M_r$ 30,000) from each of the main subunits of NR, but the remainder of the molecule and its tetrameric arrangement is not altered. Trypsin and other proteinases tested on *Chlorella* NR did not exhibit the specific and limited proteolytic cleavage demonstrated with the maize root proteinase.

The maize root proteinase increases in activity with root age[47] to become the predominant proteinase species. Its appearance is independent of $NO_3^-$ or NR. It is present in the apical region of the root in an inactive form,[48] but no significant quantity has been detected in the maize leaf.[47] Compared to trypsin, it is considerably more active on NR than a number of other proteins.[42] General proteinase substrates such as hemoglobin, casein, and azocasein are hydrolyzed,[40] and its pH optimum on these is 4.0, 6.5, and 9.0, respectively. Change in pH could, however, be influencing the sensitivity of the different protein substrates to hydrolysis as well as the proteinase itself. Maximum inactivation of NR was observed at pH 7.0.[22]

## B. Barley Leaf Proteinase

Extensive studies have been undertaken on CR species apparently released in vitro from the NR complex in the barley leaf (*Hordeum vulgare* L.),[49] especially in older leaves.[50] The $M_r$ 61,000 and 40,000 CR species were considered to represent domains cleaved from the main subunit of NR.[50,51] Inhibition of this breakdown of NR by leupeptin[52] suggests that the proteinase responsible is a cysteine proteinase.

In barley leaf extracts, loss of NR activity was correlated with the appearance of smaller CR species[50] (also demonstrated with trypsin treatment). This contrasts with the action of the maize root proteinase where the CR-part of the NR complex[42] and also free CR species were especially vulnerable to attack.[53] It has been confirmed that when trypsin inactivates maize scutella NR, there is an increase in the amount of the smaller CR species.[42] A $M_r$ 38,000 CR species in the barley leaf extracts has been characterized[54] and shown not to contain heme. Preliminary studies indicate that this CR species cross reacts with antibody to the barley leaf NR.[55]

Two cysteine proteinases have been characterized in barley leaves. One in *Hordeum distichum* L. ($M_r$ 74,000) has been shown by Hamano et al.[56] to result in loss of NADH-NR activity, but have only a slight effect on the NADH-CR component. $FMNH_2$- and MVH-NR activity were lost. The NADH-CR component, while still retaining activity, could have been cleaved from the NR complex as described by Brown et al.[50] (see above). A smaller cysteine proteinase ($M_r$ 28,300) has been isolated from the leaves of *Hordeum vulgare* L. by Miller and Huffaker,[57] but its effect on NR was not tested. It has been shown to occur in the vacuole.[58]

In contrast to the above observations, which suggest that CR species with molecular weights less than 100,000 are breakdown products of the main NR subunit, it has been claimed that a diaphorase ($M_r$ 60,000) and other low-molecular-weight species in a highly purified sample of barley leaf NR were not related to NR.[59] Analysis of peptide maps indicated no homology between the NR subunit and the smaller components. Further, only a 4S CR species ($M_r$ 110,000), the NR subunit, was shown to cross react with the NR antibody.[60] In these studies, however, precautions were taken to minimize proteolysis of NR, and the diaphorase component ($M_r$ 60,000) may be a constitutive CR species that possibly could have been removed by high speed centrifugation.[36]

## V. EVALUATION OF THE ROLE OF PROTEINASES IN THE REGULATION OF NITRATE REDUCTASE

A high rate of degradation of an enzyme in vivo and sensitivity to proteolysis in vitro appear related to some aspect of its structure or three dimensional conformation. For example, a large subunit molecular weight, the degree of hydrophobicity, a low isoelectric point or perhaps a proteinase sensitive region could account for the specific cleavage.[61,62] Indeed NR with its relatively large subunit molecular weight (Section II.A) and low isoelectric-point[63] fits with the general correlation between such properties and a high rate of turnover for animal and plant enzymes.[61,62] The similarity of half-life values reported for NR in different plants (Section II.B) and the same rate of degradation, whether in induction-phase, steady state, or decreasing enzyme level,[15] suggests a constant turnover of the enzyme and the involvement of constitutive proteinase(s) in its degradation.

Two proteinases (maize root and barley leaf) have been described that mediate a relatively specific and limited initial proteolysis of NR, possibly at sensitive hinge regions between the main domains of the enzyme.[50] The location of these proteinases in the cell needs to be ascertained since NR is located in the cytosol of the plant cell, and normal protein turnover in animal and plant cells is considered to involve a nonlysosomal route.[64] The serine proteinase, which has been characterized in the maize root, appears to be the main proteinase in this tissue. Unlike NR, which can be induced in all plant tissues, this serine proteinase has a limited distribution. Its activity is not influenced by the level of nitrate supplied or amount of NR present, but increases with root age. Thus while the barley leaf and maize root proteinase have been most helpful in illustrating the basis of the sensitivity of NR to proteolysis, they may not have a role in the initiation of degradation of the enzyme in vivo.

The initial signal for the degradation of NR is not known, perhaps a change in conformation exposing a sensitive peptide bond region. When *Chlorella* NR is reduced (reversibly inhibited), it is resistant to trypsin,[65] but its susceptibility to the maize root proteinase was not altered. When wheat leaf NR was converted to a reduced state, its sensitivity to proteolysis was not altered.[66]

A more comprehensive study is required on the spectrum of proteinases present in the plant cell, in particular in the cytosol, to elucidate the steps in the degradation of NR. Perhaps inactivator molecules (defined in Section III.A) have proteinase activity associated with them that has not been detected by assay procedures largely devised for animal proteinases. One is encouraged by the recent studies on the yeast cell,[67] where in addition to the main proteinases characterized earlier, many new proteinase species have recently been detected.

## REFERENCES

1. **Beevers, L. and Hageman, R. H.,** Uptake and reduction of nitrate: bacteria and higher plants, in *Inorganic Plant Nutrition,* Lauchli, A. and Bieleski, R. L., Eds., *Encyclopedia of Plant Physiology,* Vol. 15A (New Series), Pirson, A. and Zimmerman, M. H., Eds., Springer-Verlag, Berlin, 1983, 351.
2. **Kuo, T-M., Kleinhofs, A., and Warner, R. L.,** Purification and partial characterization of nitrate reductase from barley leaves, *Plant Sci. Lett.,* 17, 371, 1980.
3. **Kuo, T-M., Warner, R. L., and Kleinhofs, A.,** *In vitro* stability of nitrate reductase from barley leaves, *Phytochemistry,* 21, 531, 1982.
4. **Somers, D. A., Kuo, T-M., Kleinhofs, A., Warner, R. L., and Oaks, A.,** Synthesis and degradation of barley nitrate reductase, *Plant Physiol.,* 72, 949, 1983.
5. **Somers, D. A., Kuo, T-M., Kleinhofs, A., and Warner, R. L.,** Nitrate reductase-deficient mutants in barley. Immunoelectrophoretic characterization, *Plant Physiol.,* 71, 145, 1983.
6. **Fido, R. J. and Notton, B. A.,** Spinach nitrate reductase: further purification and removal of "nicked" subunits by affinity chromatography, *Plant Sci. Lett.,* 37, 87, 1984.

7. **Redinbaugh, M. B. and Campbell, W. H.**, Quaternary structure and composition of squash NADH: nitrate reductase, *J. Biol. Chem.*, 260, 3380, 1985.
8. **Nakagawa, H., Poulle, M., and Oaks, A.**, Characterization of nitrate reductase from corn leaves (*Zea mays* cv W64A × W182E). Two molecular forms of enzyme, *Plant Physiol.*, 75, 285, 1984.
9. **Pan, S-S. and Nason, A.**, Purification and characterization of homogeneous assimilatory reduced nicotinamide adenine dinucleotide phosphate-nitrate reductase from *Neurospora crassa*, *Biochim. Biophys. Acta*, 523, 297, 1978.
10. **Johnson, J. L., Hainline, B. E., and Rajagopalan, K. V.**, Characterization of the molybdenum cofactor of sulfite oxidase, xanthine oxidase and nitrate reductase, *J. Biol. Chem.*, 255, 1783, 1980.
11. **Horner, R. D.**, Purification and comparison of nit-1 and wild type NADPH: nitrate reductase from *Neurospora crassa*, *Biochim. Biophys. Acta*, 744, 7, 1983.
12. **Howard, W. D. and Solomonson, L. P.**, Quaternary structure of assimilatory NADH-nitrate reductase from *Chlorella*, *J. Biol. Chem.*, 257, 10243, 1982.
13. **Nakagawa, H., Yonemura, Y., Yamamoto, H., Sato, T., Ogura, N., and Sato, R.**, Spinach nitrate reductase. Purification, molecular weight and subunit composition, *Plant Physiol.*, 77, 124, 1985.
14. **Wray, J. L. and Filner, P.**, Structural and functional relationships of enzyme activities induced by nitrate in barley, *Biochem. J.*, 119, 715, 1970.
15. **Zielke, H. R. and Filner, P.**, Synthesis and turnover of nitrate reductase induced by nitrate in cultured tobacco cells, *J. Biol. Chem.*, 246, 1772, 1971.
16. **Mendel, R. R., Alikulov, Z. A., and Muller, A. J.**, Molybdenum cofactor in nitrate reductase-deficient mutants. III. Induction of cofactor synthesis by nitrate, *Plant Sci. Lett.*, 27, 95, 1982.
17. **Funkhouser, E. A., Shen, T-C., and Ackerman, R.**, Synthesis of nitrate reductase in *Chlorella*. I. Evidence for an inactive protein precursor, *Plant Physiol.*, 65, 939, 1980.
18. **Oaks, A., Wallace, W., and Stevens, D.**, Synthesis and turnover of nitrate reductase in corn roots, *Plant Physiol.*, 50, 649, 1972.
19. **Radin, J. W.**, Distribution and development of nitrate reductase activity in germinating cotton seedlings, *Plant Physiol.*, 53, 458, 1974.
20. **Aslam, M. and Oaks, A.**, Comparative studies on the induction and inactivation of nitrate reductase in corn roots and leaves, *Plant Physiol.*, 57, 572, 1976.
21. **Acton, G. J. and Gupta, S.**, A relationship between protein degradation rates *in vivo*, isoelectric points, and molecular weights obtained by using density labelling, *Biochem. J.*, 184, 367, 1979.
22. **Wallace, W.**, Purification and properties of a nitrate reductase-inactivating enzyme, *Biochim. Biophys. Acta*, 341, 265, 1974.
23. **Yamaya, T. and Ohira, K.**, Purification and properties of a nitrate reductase inactivating factor from rice cells in suspension culture, *Plant Cell. Physiol.*, 18, 915, 1977.
24. **Jolly, S. O. and Tolbert, N. E.**, NADH-nitrate reductase inhibitor from soybean leaves, *Plant Physiol.*, 62, 197, 1978.
25. **Yamaya, T., Oaks, A., and Boesel, I. L.**, Characteristics of nitrate reductase inactivating proteins obtained from corn roots and rice cell cultures, *Plant Physiol.*, 65, 141, 1980.
26. **Yamaya, T., Solomonson, L. P., and Oaks, A.**, Action of corn and rice-inactivating proteins on a purified nitrate reductase from *Chlorella vulgaris*, *Plant Physiol.*, 65, 146, 1980.
27. **Yamaya, T. and Ohira, K.**, Reversible inactivation of nitrate reductase by its inactivating factor from rice cells in suspension culture, *Plant Cell Physiol.*, 19, 1085, 1978.
28. **Leong, C. C. and Shen, T-C.**, Action kinetics of the inhibition of nitrate reductase of rice plants, *Biochim. Biophys. Acta*, 703, 129, 1982.
29. **Sherrard, J. H., Kennedy, J. A., and Dalling, M. J.**, *In vitro* stability of nitrate reductase from wheat leaves. III. Isolation and partial characterization of a nitrate reductase-inactivating factor, *Plant Physiol.*, 64, 640, 1979.
30. **Sherrard, J. H., Kennedy, J. A., and Dalling, M. J.**, *In vitro* stability of nitrate reductase from wheat leaves. II. Isolation of factors from crude extracts which affect stability of highly purified nitrate reductase, *Plant Physiol.*, 64, 439, 1979.
31. **Purvis, A. C., Tischler, C. R., and Funkhouser, E. A.**, The nitrate reductase stabilizing factor in cotton seeds, *Physiol. Plant.*, 48, 389, 1980.
32. **Schrader, L. E., Cataldo, D. A., and Peterson, D. M.**, Use of protein in extraction and stabilization of nitrate reductase, *Plant Physiol.*, 53, 688, 1974.
33. **Yamaya, T. and Oaks, A.**, Activation of nitrate reductase by extracts from corn scutella, *Plant Physiol.*, 66, 212, 1980.
34. **Aryan, A. P., Batt, R. G., and Wallace, W.**, Reversible inactivation of nitrate reductase by NADH and the occurrence of partially inactive enzyme in the wheat leaf, *Plant Physiol.*, 71, 582, 1983.
35. **Vennesland, B. and Guerrero, M. G.**, Reduction of nitrate and nitrite, in *Encyclopedia of Plant Physiology*, Vol. 6, Pirson, A. and Zimmerman, M. H., Eds., Springer-Verlag, Heidelberg, 1979, 425.
36. **Wallace, W.**, Effects of a nitrate reductase inactivating enzyme and NAD(P)H on the nitrate reductase from higher plants and *Neurospora*, *Biochim. Biophys. Acta*, 377, 239, 1975.

37. **Aparicio, P. J., Roldan, J. M., and Calero, F.,** Blue light photoreactivation of nitrate reductase from green algae and higher plants, *Biochem. Biophys. Res. Commun.,* 70, 1071, 1976.
38. **Streit, L. and Feller, E.,** Inactivation of N-assimilating enzymes and proteolytic activities in wheat leaf extracts: effect of pyridine nucleotides and of adenylates, *Experientia,* 38, 1176, 1982.
39. **Datta, N., Rao, L. V. M., Guha-Mukherjee, S., and Sopory, S. K.,** Activation and stabilization of nitrate reductase by NADH in wheat and maize, *Phytochemistry,* 22, 821, 1983.
40. **Shannon, J. D. and Wallace, W.,** Isolation and characterisation of peptide hydrolases from the maize root, *Eur. J. Biochem.,* 102, 399, 1979.
41. **Smith, S. C.,** Partial Purification of Nitrate Reductase Inactivators from Maize Roots, Thesis, University of St. Andrews, United Kingdom, 1982.
42. **Batt, R. G. and Wallace, W.,** A comparison of the effect of trypsin and a maize root proteinase on nitrate reductase and other enzymes from maize, *Biochim. Biophys. Acta,* 744, 205, 1983.
43. **He, W-z, Zhang, D-y., and Tang, Y-w.,** Studies on nitrate reductase. III. Specificity of nitrate reductase inactivating protein and its proteolytic action on nitrate reductase, *Acta Physiol. Sin.,* 9, 151, 1983.
44. **Poulle, M., Bzonek, P., and Oaks, A.,** Characterization of nitrate reductase in corn leaves ( *Zea mays* W64A × W182E): subunit size and sensitivity to the corn root protease, Personal communication, 1984.
45. **Khan, F. R., Saleemuddin, M., Siddiqi, M., and McFadden, B. A.,** The appearance and decline of isocitrate lyase in flax seedlings, *J. Biol. Chem.,* 254, 6938, 1979.
46. **Solomonson, L. P., Howard, W. D., Oaks, A., and Yamaya, T.,** Mode of action of natural inactivator proteins from corn and rice on a purified assimilatory nitrate reductase, *Arch. Biochem. Biophys.,* 283, 469, 1984.
47. **Wallace, W. and Shannon, J. D.,** Proteolytic activity and nitrate reductase inactivation in maize seedlings, *Aust. J. Plant Physiol.,* 8, 211, 1981.
48. **Wallace, W.,** Proteolytic inactivation of enzymes, in *Regulation of Enzyme Synthesis and Activity in Higher Plants,* Smith, H., Ed., Academic Press, London, 1978, 177.
49. **Small, I. S. and Wray, J. L.,** NADH nitrate reductase and related cytochrome *c* reductase species in barley, *Phytochemistry,* 19, 387, 1980.
50. **Brown, J., Small, I. S., and Wray, J. L.,** Age dependent conversion of nitrate reductase to cytochrome *c* reductase species in barley leaf extracts, *Phytochemistry,* 20, 389, 1981.
51. **Campbell, J. McA. and Wray, J. L.,** Purification of barley nitrate reductase and demonstration of nicked subunits, *Phytochemistry,* 22, 2375, 1983.
52. **Wray, J. L. and Kirk, D. W.,** Inhibition of NADH-nitrate reductase degradation in barley leaf extracts by leupeptin, *Plant Sci. Lett.,* 23, 207, 1981.
53. **Wallace, W. and Johnson, C. B.,** Nitrate reductase and soluble cytochrome *c* reductase(s) in higher plants, *Plant Physiol.,* 61, 748, 1978.
54. **Campbell, J. McA., Small, I. S., and Wray, J. L.,** Purification of a NADH cytochrome *c* reductase species from cell-free extracts of barley, *Phytochemistry,* 23, 1391, 1984.
55. **Campbell, J. McA. and Wray, J. L.,** Personal communication, 1984.
56. **Hamano, T., Oji, Y., Okamoto, S., Mitsuhashi, Y., and Matsuki, Y.,** Purification and characterization of thiol proteinase as a nitrate reductase-inactivating factor from leaves of *Hordeum distichum* L., *Plant Cell. Physiol.,* 25, 419, 1984.
57. **Miller, B. L. and Huffaker, R. C.,** Partial purification and characterization of endoproteinases from senescing barley leaves, *Plant Physiol.,* 68, 930, 1981.
58. **Thayer, S. S. and Huffaker, R. C.,** Vacuolar localization of endoproteinases $EP_1$ and $EP_2$ in barley mesophyll cells, *Plant Physiol.,* 72S, 118, 1983.
59. **Kuo, T-M., Somers, D. A., Kleinhofs, A., and Warner, R. L.,** NADH-nitrate reductase in barley leaves. Identification and amino acid composition of subunit protein, *Biochim. Biophys. Acta,* 708, 75, 1982.
60. **Narayanan, K. R., Somers, D. A., Kleinhofs, A., and Warner, R. L.,** Nature of cytochrome *c* reductase in nitrate reductase-deficient mutants in barley, *Mol. Gen. Genet.,* 190, 222, 1983.
61. **Goldberg, A. L. and St. John, A. C.,** Intracellular protein degradation in mammalian and bacterial cells. II, *Annu. Rev. Biochem.,* 45, 747, 1976.
62. **Coates, J. B. and Davies, D. D.,** The molecular basis of the selectivity of protein degradation in stressed senescent barley (*Hordeum vulgare* cv. Proctor) leaves, *Planta,* 158, 550, 1983.
63. **Notton, B. A., Hewitt, E. J., and Fielding, A. H.,** Isoelectric focussing of spinach nitrate reductase and its tungsten analogue, *Phytochemistry,* 11, 2447, 1972.
64. **Ballard, F. J.,** Intracellular protein degradation. *Essays Biochem.,* 13, 1, 1977.
65. **Howard, W. D. and Solomonson, L. P.,** Trypic cleavage of nitrate reductase, Proc. 176th Natl. Am. Chem. Soc. Meet., Abstr. 89, 1981.
66. **Aryan, A. P. and Wallace, W.,** Action of a maize root proteinase and trypsin on nitrate reductase from maize scutellum and wheat leaf, *Plant Physiol.,* 72S, 110, 1983.
67. **Wolf, D. H.,** Proteinase action *in vitro* versus proteinase function *in vivo*: mutants shed light on intracellular proteolysis in yeast, *Trends Biochem. Sci.,* 7, 35, 1982.

Chapter 6

# A MEMBRANE-DERIVED PROTEINASE CAPABLE OF ACTIVATING A GALACTOSYL-TRANSFERASE INVOLVED IN VOLUME REGULATION OF *POTERIOOCHROMONAS*\*

**Heinrich Kauss**

## TABLE OF CONTENTS

I. Introduction: Regulation of Cell Volume and Turgor Pressure ................... 92

II. Physiology of Isofloridoside Metabolism in *Poterioochromonas* ................ 93

III. Purification and Properties of the Activating Proteinase ......................... 96

IV. $Ca^{2+}$-Dependent Generation of the Proteinase from Membranes ................ 97

V. Influence of Cell Shrinkage on Proteinase Activity in Crude Homogenates and its Physiological Significance ..................................................... 98

Acknowledgments ............................................................................ 101

References ..................................................................................... 101

---

\* *Abbreviations:* IF, isofloridoside or α-galactosyl-(1 → 1)-glycerol; IFP, isofloridoside-phosphate or α-galactosyl-(1 → 1)-glycerol-3-phosphoric acid; IFP-synthase, UDPGal: *sn*-glycerol-3-phosphoric acid 1-α-D-galactosyl-transferase (EC 2.4.1.96).

## I. INTRODUCTION: REGULATION OF CELL VOLUME AND TURGOR PRESSURE

Plasma membranes have limited permeability to most substances solubilized in cells, but they are greatly permeable to water. Cells, therefore, represent an osmotic system into which water will enter as long as the osmotic potential outside is less negative than inside. This results in the build-up of a turgor pressure in walled cells of plants or bacteria, accompanied by a relatively small volume increase. In contrast, naked cells such as plant gametes, protists, or animal cells would show a dramatic volume increase and a rather small turgor. Physiological studies have provided many examples which show that either turgor pressure or cell volume can be regulated by alteration of the concentration of cell solutes.[1,2,3] On the one hand, such a regulation of the cellular water content can serve to either hold volume or turgor constant when the external osmotic value varies, for instance, in algae. These systems, however, operate not only if required by changing ecological situations, but also when the cell volume increases due to cell growth, or when certain cellular solutes are used up or formed to fulfill metabolic necessities. On the other hand, there are many cells or tissues of special design that can perform work by means of increasing the solute concentration that enforces a water inflow. Stomata or leaf pulvini provide well-known examples.

The solutes involved in these systems are in many cases inorganic ions, mainly $K^+$. Alternatively, low-molecular-weight organic molecules occur that might function either directly as osmotic agents, or serve in addition also as counterions. Examples include sugars, sugar alcohols, malate, and amino acids as well as such unusual substances as cyclohexanetetrol and isofloridoside ($\alpha$-galactosyl-$(1 \rightarrow 1)$-glycerol).[1,2,3] Recognition of the role of a certain substance as an osmotic solute is often occluded by the fact that in one type of cell several substances are used in parallel or possibly in different compartments, and that in addition to an osmotic purpose, certain compounds (proline, betaine, glycerol) may also serve to stabilize critical membrane proteins under conditions of more severe water stress.[1,4]

In any case, where the cellular concentration of a solute has to be changed in order to provide a means of water regulation, a respective mechanism has to receive some type of information as to the degree and direction of the desired change. In the case of inorganic ions, the regulation might occur at transport proteins; however, at a molecular level, regulation of such "pumps" can hardly yet be understood. In systems where organic molecules are produced or used up to affect the osmotic balance, the regulation of the relevant metabolic enzymes is more likely to be accessible to experimental analysis. Any regulative mechanism requires that either the turgor pressure or a parameter related to cell volume is measured, and a signal capable of steering the relevant biochemical process is derived from such a measurement. Some investigators have tried to apply principles of system theory and to construct feedback loops in order to elucidate the problems involved.[1,2,5] Such attempts might help to clarify the elements of the processes involved, but this has remained fully abstract in most details up to now. Long-term potassium transport in the green alga *Valonia* provides the only case where experimental evidence is available to promote an idea as to the nature of the pressure-sensing mechanism.[5,6] It has been found that a gradient in hydrostatic pressure over the plasma membrane/wall complex is likely to provide the information. This has led to the suggestion that bending and/or compression of the membranes might be the critical initial event.[5,6]

In the golden-brown, wall-less unicellular alga *Poterioochromonas malhamensis* (syn. *Ochromonas malhamensis*) the major osmotic solute is isofloridoside (IF). Experimental results accumulating over many years have led to the hypothesis (Figure 1) that during shrinkage induced by an increase in external osmotic value, the signal resulting from a decrease in volume and/or cell surface may, in a membrane-located reaction sequence, be converted into activation of a specific proteinase that is capable of activating the enzyme

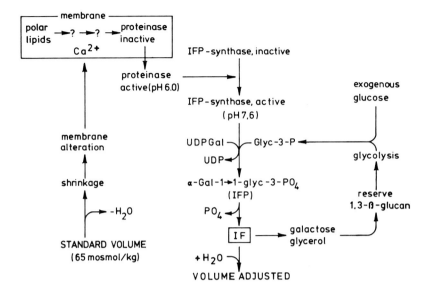

FIGURE 1. Isofloridoside metabolism in *Poterioochromonas* and proposed events leading to proteolytic activation of IFP-synthase as a consequence of cell shrinkage. For explanation see text.

α-galactosyl-(1 → 1)-glycerol-3-phosphoric acid (IFP) synthase. This cascade-like mechanism is suggested to be one of the possibilities able to render this key enzyme of IF metabolism active as a consequence of cell shrinkage. Many details of the suggested mechanism remain to be elaborated upon, and it is also not clear as to which extent similar mechanisms may operate in osmotic regulation by other organisms or cells. Nevertheless, the reactions described provide, in addition to their significance in cell water regulation of *Poterioochromonas*, an example of how a physiological stimulus might be transmitted into physiological action by means of limited proteolysis. This appears of special interest in the context of the increasing number of reports[7,8] on apparently specific proteinases for which often the function, and in most cases also, the way they are brought to action remains unknown.

## II. PHYSIOLOGY OF ISOFLORIDOSIDE METABOLISM IN *POTERIOOCHROMONAS*

Addition of solutes such as sucrose, mannitol, or sodium chloride to a cell suspension of *Poterioochromonas* causes the cells to shrink within about 1 min. The volume attained is indirectly proportional to the osmotic pressure applied. An increase from 65 to 180 mosmol/kg, for instance, reduces the volume by about 20%. The lost water is regained nearly quantitatively within 1 to 2 hr through an internal accumulation of solutes. Determination of cell volume and its osmotic portion as well as the nature and amount of selected cellular solutes showed that 70 to 80% of the observed volume regained was due to the biosynthesis of IF. Another 10% can be accounted for by the increase in free amino acids (mainly alanine), and about 10 to 20% are contributed by $K^+$ and its unknown counterion.[1,2] IF-metabolism, therefore, is the major producer of osmotic solutes, and for the reasons indicated above, research was focused on the enzymes involved.

Some features of the IF-pathway were already evident from in vivo studies. About 1 to 3 min after addition of an osmotic agent, IF is produced at a constant rate. This, together with studies using inhibitors of protein synthesis, suggested that this initial part of the overall regulation does not involve *de novo* synthesis of an enzyme, but may occur instead by activation of a preexisting enzyme. Chase experiments with [$^{14}$C]-glucose indicated that the rate of IF-formation was about tenfold higher (cf., control) shortly after shrinkage.[9] However, by the time the cells had regained their original volume, as a consequence of IF accumulation, the rate of IF production was only half this rate, and net accumulation of IF had stopped. This indicates a rapid turnover of IF. When the external osmotic pressure was reduced, production of IF stopped immediately, and degradation of IF was enhanced about threefold. Under these conditions, all the carbon from the galactose and glycerol moiety of IF was recovered in the reserve 1,3-β-D-glucan, chrysolaminarin. This suggests that the pool size of IF is regulated at the site of production as well as at the site of degradation, and that the low-molecular-weight osmotic substance IF is in reversible exchange with the reserve glucan. This polymer exhibits a molecular weight of about 6000 and is located in a storage vacuole, which, in the young cells routinely used, fills a prominent part of the cell volume.

Evidence has accumulated that shows that several enzymes involved in the IF pathway (Figure 1) may be under control. As described above, IF degradation increased immediately after the external osmotic pressure was reduced by addition of water. In addition to the short-term regulation evident from these in vivo experiments, the α-galactosidase necessary to remove the α-linked galactose from IF shows increased in vitro activity in cells that have regained their original volume 1 to 2 hr after the initial shrinkage event.[10] Similar observations were made for the galactokinase.[11] During experiments with cells prelabeled by photosynthesis with $^{14}CO_2$, we observed that an increase in volume due to dilution of the suspension fluid could induce first an increase, followed by a decrease, of the IFP pool size.[12] This unexpected time course might indicate that the IFP-phosphatase is an additional site of regulation. Although this enzyme has been purified to some extent,[13] we have not yet identified any relevant regulatory properties. Vice versa, the phosphorylase responsible for degradation of the 1,3-β-D-glucan has been shown to be activated by some adenylates,[14] but it remains unclear whether this is of physiological significance.

Our more recent studies focused on the IFP-synthase because the flow of carbon through this enzyme obviously accelerated rapidly shortly after cell shrinkage occurred. In crude extracts from cells which were shrunken for a few minutes, we observed an increase in apparent activity that was proportional to the degree of shrinkage (see also zero time values of Figure 2) and that was nullified a few minutes after cells were artificially reswollen.[15] We could demonstrate this change in activity of the IFP-synthase despite the fact that the assay procedure for IFP-synthase results in a 1000-fold dilution of the cell constituents. Accordingly, any potential allosteric effector involved would have been greatly diluted. Therefore, and in agreement with the rapid onset of IF-synthesis in vivo, at least one way to regulate the IFP-synthase during (or shortly after) cell shrinkage might be by way of a covalent modification of unknown nature.

Studies of the in vivo IFP-synthase activation in relation to time indicated that the process is complex. There is an initial period of 5 to 10 min in which the enzyme activity, as indicated above, reflects the degree of shrinkage. After about 1 hr a time when the IF pool is already filled to 75% and the rate of IF synthesis tends to be diminished, a second peak of even higher in vitro activity of IFP-synthase can be observed.[15] This discrepancy can only be explained by the assumption that the enzyme activity might be diminished in the cell by a superimposed negative regulation principle of unknown nature. The IFP-synthase found to be very active 1 hr after shrinkage voids a Sephadex® G 200 column and thus appears to have a molecular weight ≥450,000, whereas the apparently inactive form from unshrunken cells elutes from the column with a molecular weight of 100,000 to 160,000. The activation process, therefore, appears to be accompanied by, or result from, aggregation. The appearance of active IFP-synthase 1 hr after onset of shrinkage is a transient event. This type of active enzyme has mostly disappeared again 1 hr later, a time at which the IF-pool is high and remains of constant size, but is under rapid turnover.[15]

As our major interest was to contribute towards the understanding of the signal transmitting

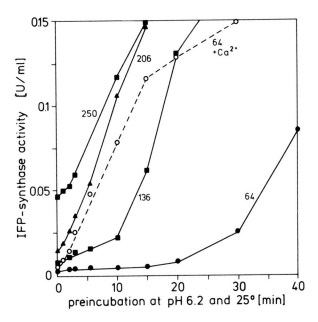

FIGURE 2. Time course of IFP-synthase activation in crude homogenates from cells shrunken to various degrees. Cells of standard volume (●——●); 64 mosmol/kg) or cells shrunken for 2 min by addition of sucrose solution to give 136 (■——■), 206 (▲——▲) or 250 (▼——▼) mosmol/kg were homogenized at zero time and the homogenates incubated at pH 6.1 and 25°C. Aliquots were withdrawn at the times indicated and IFP-synthase assayed immediately at pH 7.6 for 5 min. Part of the standard volume cell homogenate (○ - ○) was brought to about 200 μM free $Ca^{2+}$ by addition of $CaCl_2$ in amounts sufficient to titrate the EDTA present in the homogenate. One unit of IFP-synthase produces 1 μmol of $^{14}C$-IFP/min.[20] The values at zero time reflect the IFP-synthase present in the cells. One unit of activating enzyme ("proteinase") is defined as that activity which increases the IFP-synthase activity by one unit per minute.[20] Its activity was calculated from the initial slope of the curves and was found to be 0.9, 1.6, 6.6, and 4.1 mU/mℓ for the cells maintained at 64, 136, 206, 250 mosmol/kg, respectively. Addition of $Ca^{2+}$ to the homogenate from standard cells increased it from 0.9 to 6.1 mU/mℓ. For a more detailed explanation of the spontaneous increase of activation (slope of the curves) on prolonged incubation, see the text. (Data taken from Reference 21.)

step, we focused our further work on the first few minutes after shrinkage. It was found that IFP-synthase in homogenates from standard volume cells (external osmotic pressure 65 mosmol/kg) showed very low activity, but could be activated by incubation with trypsin or chymotrypsin.[16] In addition, incubation of the crude cell homogenate prepared as indicated above resulted in an activation process that was dependent on the pH value (apparent optimum around 6.2), time, and temperature and did not require the addition of nucleotides.[17,18] Taken together, these results suggested that activation in vitro occurs due to an endogenous auxiliary enzyme, which possibly might be of proteinase nature.

During the above studies, and over several years, we tried to optimize the conditions of cell disintegration with the goal of reflecting the physiological situation, namely to obtain homogenates from cells of standard volume that showed IFP-synthase activity as low as possible. To this end, it was found to be important to homogenize the cells at a certain suspension density in a Yeda-press (not with a sonifier!) in the presence of 2 m$M$ EDTA or EGTA, bovine serum albumin, freshly added 2-mercaptoethanol, and in 100 m$M$ MES/NaOH pH 6.2. It will be discussed later (Sections IV and V) that under these conditions not only the IFP-synthase activity was low, but also the activating mechanism (Figure 1) was not triggered (or only to a relatively small extent). If no care is taken to control the parameters mentioned, then rapid activation of IFP-synthase occurs during and within a short time after homogenization. As this proceeds even at low temperatures, serious problems can arise for the evaluation of the physiological significance of properties found for the IFP-synthase purified from conventionally homogenized cells.[19] These problems have also to be considered in regard to the question of whether the inactive IFP-synthase represents a zymogen and whether this presumed precursor undergoes molecular weight changes on induction of activation by cell shrinkage.

## III. PURIFICATION AND PROPERTIES OF THE ACTIVATING PROTEINASE

As indicated in Section II, IFP-synthase activation in vitro did not occur when cells of standard volume were homogenized under certain precautions. We had also observed that the activating enzyme was active and partly soluble when cells were homogenized in a 10 m$M$ buffer instead of the 100 m$M$ routinely used. This observation greatly facilitated purification of the activating enzyme. We simply eluted the activating enzyme by resuspending the crude membrane pellet in water. This treatment removed most of the soluble cellular proteins and the bovine serum albumin, which had to be added in the homogenization buffer. Water was chosen as eluent, as this left most of the other membrane proteins insoluble. The extracts prepared in this way were further purified to homogeneity by chromatography on DEAE-Sephacel and immobilized hemoglobin and fetuin.[20]

The activating enzyme was assayed during purification by measuring its ability to activate the inactive form of IFP-synthase. This latter "substrate" had to be partially purified to remove the minor amounts of activating enzyme that are always present even in supernatants from standard volume cells. The assay procedure was similar to that indicated in Figure 2 for crude homogenates. The properly diluted samples of activating enzyme were incubated at pH 6.0 and 25°C with the inactive IFP-synthase preparation, and aliquots were removed after various times for an immediate determination of IFP-synthase activity at pH 7.6 (see Figure 1). Definition of units for enzyme activity can be found in the legend of Figure 2.

The purified activating enzyme has a molecular weight of 46,500 on SDS-PAGE and of 45,000 on Sephadex® G150 gel permeation chromatography and thus consists of a single peptide chain. It has a sharp optimum at pH-values between 5.9 and 6.1, with only 10 or 50% of the maximal activity attained at pH 5.5 or 6.5, respectively. The activating enzyme was not inhibited by EDTA or EGTA, NaF, SH-reagents, leupeptin, bestatin, or various peptide trypsin inhibitors. In contrast, some inhibition was observed after preincubation with 2 m$M$ diisopropylfluorophosphate and phenylmethylsulfonylfluoride. About 50% inhibition also resulted from addition of chymostatin (2.5 µg/m$\ell$), antipain (37 µg/m$\ell$) and fetuin (100 µg/m$\ell$). From the many commercial proteins tested, only casein, dimethylcasein, and the bovine insulin B-chain turned out to be significantly inhibitory; when incubated with the activating enzyme, the latter three proteins also appear to be split to some extent, as indicated by the appearance of primary amino groups (fluorescamine assay). These observations, taken together, strongly suggest that the activating enzyme is a serine proteinase, possibly requiring a specific peptide sequence for its action. It is understandable, therefore, that the enzyme did not split any of the numerous chromogenic or fluorogenic artificial substrates used.[20] The peptide sequence required might become evident in future studies using the bovine insulin B-chain as a substrate.

In an attempt to show the action of the activating proteinase directly, we also incubated it together with the partially purified inactive IFP-synthase preparation. A certain band, corresponding to $M_r$ 75,000, was greatly diminished on SDS-PAGE. This observation again shows that the activating enzyme might be a proteinase. Unfortunately the identity of the band that diminishes remains unknown, as well as the position of the degradation products. This is mainly due to the fact that the inactive IFP-synthase preparation available up to now is still rather impure. It gives several bands in the region where the purified monomer of the active IFP-synthase ($M_r$ about 70,000)[19] has to be assumed; the products of the proteinase reaction may thus be hidden under one of these impurities. It thus remains to be clarified by further research whether or not the inactive IFP-synthase indeed represents a zymogen, as proposed in Figure 1.

Some indirect evidence suggests that in addition to activation by limited proteolysis, the IFP-synthase might also be activated by alternative mechanisms that in vitro can be mimicked by the presence in the IFP-synthase assay mixture of chelating ions at extremely high concentration[15] (50 m$M$ sodium pyrophosphate or 100 m$M$ EDTA).

## IV. $Ca^{2+}$-DEPENDENT GENERATION OF THE PROTEINASE FROM MEMBRANES

As indicated above (Section II), we have optimized the conditions for disintegration of standard volume cells to prepare homogenates exhibiting low IFP-synthase and activating proteinase activity. When such extracts were supplied with $CaCl_2$ in amounts sufficient to titrate the EDTA or EGTA present in the homogenization buffer, and to raise the concentration of free $Ca^{2+}$-ions into the μ$M$ range, an activating enzyme is detected that effects IFP-synthase activation to about the same degree as optimal cell shrinkage (Figure 2).[18,20,21] Most importantly, when the $Ca^{2+}$ ions were complexed again by EDTA 1 min after they were added, the process could not be stopped. This suggested that the activating enzyme itself is not stimulated by $Ca^{2+}$, but is the output of an irreversible process initiated by $Ca^{2+}$. This is in agreement with the more recent finding (Section III) that the activating proteinase itself is not $Ca^{2+}$-dependent. That the activating enzyme generated in the crude homogenate by $Ca^{2+}$ is indeed identical, or similar to, the activating proteinase described in Section III, has now been shown using chymostatin and antipain as inhibitors.[20]

Generation of the activating proteinase by $Ca^{2+}$ required the presence of membranes, and it was also shown that it became partly soluble during this process.[21] The nature of the membranes involved is still unclear. However, about two thirds of the necessary material can be spun out at rather low speed (1 min, Eppendorf bench-top centrifuge, about 10,000 g) and thus might not consist of small vesicles but rather of large membrane complexes.[22] In chrysomonades the nucleolar membrane is connected with the ER, which additionally surrounds the chloroplasts. All these membranes tend to stay together after cell homogenization. One can see with the phase contrast microscope that in the homogenates such complexes occur and some extended sheet-like structures — which might be derived from the plasma membrane — adhere. It is tempting to speculate that the latter may be the source of the activating proteinase (see Figure 1). We hope to show this in the future using antibodies against the now purified enzyme.

The generation of the activating enzyme by $Ca^{2+}$ was shown to be enhanced up to threefold upon the addition of calmodulin.[21,22] Both the calmodulin isolated from *Poterioochromonas* or from bovine brain were equally effective. This was taken as an indication that the reaction sequence operating in membranes (Figure 1) recognizes the $Ca^{2+}$ added as a $Ca^{2+}$/calmodulin complex. It appears of interest and possibly is of future heuristic value that the calmodulin-enhanced $Ca^{2+}$ effects are more rapid in homogenates prepared in the presence of fluoride or molybdate ions.[22] These substances are phosphatase inhibitors. The effect might, therefore, indicate participation of a phosphorylated compound whose dephosphorylation during cell homogenization is slowed down. This could indicate that one of the unknown membrane-located intermediates (? in Figure 1) in the sequence generating the activating proteinase is a phosphorylated compound.

It was more difficult to demonstrate that the effect which results from addition of $Ca^{2+}$ alone to crude cell homogenates is due to endogenous calmodulin. We first tried to show this with the help of the calmodulin-binding substances trifluoperazine, fluphenazine, and chlorpromazine.[18] These phenothiazine drugs can bind to calmodulin only in the presence of $Ca^{2+}$-ions and are often used to indicate an involvement of calmodulin in $Ca^{2+}$-dependent reactions or physiological processes. However, these substances exhibited a stimulating effect instead of inhibiting and curiously did so in the absence of $Ca^{2+}$. It soon became evident that this was due to a detergent-like action of the drugs,[23] a fact which will be discussed in more detail in Section V. An inhibition of the $Ca^{2+}$-induced proteinase generation was found[22] with calmidazolium (substance R 24571), another $Ca^{2+}$/calmodulin binding inhibitor, at concentrations above 100 $\mu M$, and this was first taken as an indication that endogenous calmodulin might mediate the effect of $Ca^{2+}$. In the meantime it became evident, however, that this substance, at the concentrations used, can also inhibit phospholipid-dependent enzymes that are $Ca^{2+}$-stimulated, but *not* calmodulin-dependent, such as the animal protein kinase C[24] and the membrane-bound 1,3-β-D-glucan synthase from soybean suspension cells.[25] Therefore, the question whether or not the effect of $Ca^{2+}$ is mediated by endogenous calmodulin present in the crude cell homogenate is once again open.

The observation that $Ca^{2+}$-ions in vitro can mimic a process triggered in living cells by shrinkage (Figure 2) suggested to us — as a working hypothesis — that changes in cytoplasmic free $Ca^{2+}$, possibly occurring as a consequence of cell shrinkage, might be a means to transmit the physiological stimulus and initiate the activation cascade (Figure 1). We have tried to find some circumstantial evidence to sustain this assumption with in vivo experiments. In chlorotetracycline-treated cells, the fluorescence signal increases to some extent on shrinkage, reaching a maximum after 3 min and declining again after 5 min. This time course and the fact that chlorotetracycline under the conditions used seems predominantly to stain $Ca^{2+}$ in the polar environment of membrane vesicles indicate that this might be due to a shrinkage-induced rearrangement of the internal $Ca^{2+}$ pools.[26] It has also been found that shrinkage-induced formation of IF appears not to require the presence of external $Ca^{2+}$, as this process also proceeds when the external $Ca^{2+}$ is complexed by EGTA. However, IF synthesis is greatly reduced under these conditions when cells were preincubated with the $Ca^{2+}/Mg^{2+}$-ionophore A 23187.[26] This might mean that the presumed increase in cytoplasmic $Ca^{2+}$ cannot occur, as the $Ca^{2+}$ liberated from internal pools is now flowing out of the cells at an increased rate and is trapped outside by the EGTA. As the overall metabolism of cellular $Ca^{2+}$ is a complex process and the methods employed are rather indirect, these results provide only preliminary indications and have also to be seen in conjunction with possible alterations in the lipid phase of membranes, as discussed below.

## V. INFLUENCE OF CELL SHRINKAGE ON PROTEINASE ACTIVITY IN CRUDE HOMOGENATES AND ITS PHYSIOLOGICAL SIGNIFICANCE

When it was recognized that the IFP-synthase in crude cell homogenates can be activated by an endogenous auxiliary enzyme, we tried to show whether or not the physiological stimulus "shrinkage" can influence the system. The initial slope of IFP-synthase activation at pH 6.1 was found to be increased in homogenates from cells shrunken for a few minutes. When cell homogenization was performed 2 min after addition of the osmotic solute (Figures 2 and 3), this increase was roughly proportional to the degree of shrinkage in the range between 65 and 200 mosmol/kg.[18,21] Most importantly, if shrunken cells were artificially

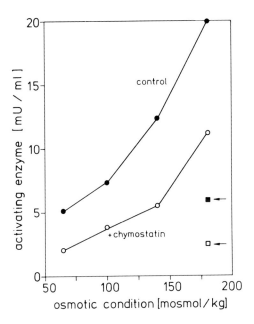

FIGURE 3. Inhibition by chymostatin of the activating enzyme in homogenates from shrunken cells as an indication of its proteinase nature. Treatment of cells and determination of activating enzyme as in Figure 2. The homogenates used are depicted as control: ●, ■; or supplied with 5 μg/mℓ of chymostatin ○,□. The arrows, ■,□ indicate values for homogenates from cells that were first shrunken for 2 min at 180 mosmol/kg followed by addition of water to lower again the osmotic value to 63 mosmol/kg and further aeration for 4 min. Similar inhibition was found by antipain (25 μg/mℓ, results not shown). (Data taken from Köhle, D. and Kauss, H., Biochim. Biophys. Acta, 799, 59, 1984.)

reswollen by addition of water, the effect was nullified a few minutes later (Figure 3). Cell shrinkage, therefore, appears to trigger the mechanism proposed to reside in membranes (Figure 1) and, if no longer required, the induced alteration is reversed.

Similarly to the activating proteinase generated by $Ca^{2+}$ (Section IV), the proteinase found to be active in homogenates from shrunken cells apparently is partly soluble. Another part is active, but can be spun out during a 5-min run on a Beckman Airfuge (about 120,000 g). These unpublished results may be taken as evidence that the proteinase during its generation first becomes active within the membrane and then, as a secondary process, becomes soluble. That the proteinase is not simply liberated from membrane-coated vesicles, but generated by a more complex process, is also indicated by the observation that sonication of the homogenates does not lead to the appearance of the proteinase but, in contrast, abolishes its generation (see also the discussion below of the results presented in Table 1).

The above results were first found with homogenates prepared in the absence of bovine serum albumin.[18] As the initial reactions involved are rather rapid, the preincubation at pH 6.1 had to be performed at 6°C under these conditions. It was then found that the induction process was slowed down by addition of bovine serum albumin to the homogenization buffer. The bovine serum albumin appears to interfere with a certain reaction occurring in the membrane (Figure 1), whereas the purified activating proteinase itself is not inhibited.[20] All further experiments could, therefore, be performed more conveniently at 25°C and resulted in similar observations.[21]

The recent recognition of very effective inhibitors for the activating proteinase[20] allows some characterization of the activating enzyme present in the crude homogenates. It is inhibited by low concentrations of chymostatin (Figure 3) and antipain[20] and thus appears indeed to be identical or similar to the proteinase purified to homogeneity (Section III). There is preliminary evidence from the pH-optimum and antipain-sensitivity that in crude homogenates more than one proteinase might be involved in the overall regulation process. In addition, we have also observed inactivation of the activating proteinase in crude homogenates.[17] One might speculate that this indicates action of other proteinases that might also be responsible for the rapid disappearance in vivo of the activating proteinase (Figure 3) as well as of the active IFP-synthase[15] shortly after artificial cell swelling.

Another feature evident from Figure 2 is that activation (slope of the curves) increases spontaneously after a certain time of preincubation. The lag period is progressively reduced with increasing cell shrinkage; with the higher osmotic treatments (e.g., 206 mosmol/kg in Figure 2), the initial slope is almost the same as that in the second phase. The spontaneous increase of activation can be prevented by centrifuging out membranes and appears to reflect an early manifestation of cell shrinkage,[18] probably due to a persistent chemical alteration of an unknown membrane constituent. The numbers given for the units of activating proteinase in crude homogenates (Figures 2 and 3), therefore, reflect only in part the situation in the cell, but in addition also some activating enzyme generated during the preincubation period. The tendency for the latter is higher for extensively shrunken cells and also shows some variation with the batch of algae used (compare the legend of Figure 2 with the data in Figure 3). This might have several reasons. The "standard volume" of cells is more or less arbitrarily defined as that volume exhibited by cells suspended at 65 mosmol/kg. Under these conditions the IF-pool is low, but not zero and still shows some turnover.[9] It thus appears understandable that in some experiments the initial slope and the onset of spontaneous activation for cells of "standard volume" is similar, for instance, to that for 136 mosmol/kg in Figure 2. Another experimental compromise is that cell harvesting requires about 2 min during which the cells remain without proper oxygen supply. It is also not fully satisfying that homogenization, although quickly performed at low temperature, has to occur in a buffer of about 120 mosmol/kg, which represents another short shrinkage or swelling period (dependent on the osmotic value of the suspension medium used). In addition to cell volume regulation exerted by fluctuations in the concentration of IF and other solutes, *Poteriochromonas* also has another system for cell water regulation, namely the pulsating vacuole, which has to be regulated cooperatively. The variations in efficiency of the IFP-synthase activation system observed in different batches of algae used, although inconvenient for the researcher, nevertheless appear intelligible.

The spontaneous onset of the activation mechanism in crude homogenates appears to be related to additional observations. During attempts to demonstrate an involvement of endogenous calmodulin in the $Ca^{2+}$-induced generation of the activating enzyme we observed[21,23] that the phenothiazine drugs used, as well as certain detergents (Triton®-X-100 and digitonin, but not octylglucoside) were also able to trigger activation in vitro in the absence of $Ca^{2+}$ (Section IV). This suggested that the reaction sequence postulated to occur in membranes (Figure 1) is somehow dependent on the general fluidity, or on the availability of certain membrane lipids. Such an assumption may be sustained by the recent finding that the spontaneous, as well as $Ca^{2+}$-induced generation of the activating proteinase, is inhibited by some membranotropic substances known to interact with polar lipids (Table 1). Polymyxin B, a basic cyclic peptide with a hydrophobic side chain, has been suggested to bind to polar lipids and to intercalate into membranes.[24,27] This antibiotic as well as the other substances used in Table 1 can inhibit, for instance, the phospholipid-dependent $Ca^{2+}$-stimulated protein kinase C.[24,27]

Our present working hypothesis thus assumes that one of the initial events of stimulus

## Table 1
### INHIBITION OF SPONTANEOUS AND $Ca^{2+}$-STIMULATED GENERATION OF THE ACTIVATING PROTEINASE BY VARIOUS PHOSPHOLIPID-INTERACTING SUBSTANCES

| | Activating proteinase[a] [mU/m$\ell$] | |
|---|---|---|
| Additions (final concentration) | Spontaneous | + $Ca^{2+}$ |
| None (control) | 5.0 | 14.2 |
| Polymyxin B sulfate (0.5 m$M$) | 1.9 (62)[b] | 2.6 (82) |
| Doxorubicin (1 m$M$) | 2.9 (42) | 2.5 (82) |
| Palmitoyl-DL-carnitine (1 m$M$) | 3.9 (22) | 5.6 (61) |

[a] Aliquots of homogenate from standard volume cells were supplied at 0°C with either of the substances or $Ca^{2+}$ and then incubated for 3 min at 25°C. The activity of the activating proteinase was measured according to the procedure outlined in the legend to Figure 2.

[b] Parenthesis: inhibition expressed as percent of control (no addition).

perception might be a change in membrane lipid density or lipid composition resulting from the shrinkage process. This idea is strongly influenced from recent developments in research with animal systems (e.g., blood platelets or mast cells) where it becomes more and more clear that in addition to changes in the concentration of cyclic AMP and $Ca^{2+}$, widely regarded to be "secondary messengers", rapid changes in the amount of membrane-bound diacylglycerol, a product of phosphatidylinositol turnover, may play a role in signal transfer.[28] The idea that membrane lipids might also be important in the signal transfer step during cell shrinkage of *Poterioochromonas* is not in contradiction with the remarks in Section IV regarding a possible role of $Ca^{2+}$-ions in the same process. It has been found previously that incubation of *Poterioochromonas* cells with the $Ca^{2+}$-ionophore A 23187 can partly mimic the effect of cell shrinkage on IFP-synthase activation.[18] The degree of this effect is variable to some extent and is sometimes not very pronounced. Although the respective mechanisms involved are most likely quite different, this phenomenon is reminiscent again of the above-cited animal cell systems where ionophore-induced $Ca^{2+}$-influx and membrane lipid changes also appear to cooperate synergistically in signal transfer.[21]

## ACKNOWLEDGMENTS

The work described in this report was supported by several grants from the *Deutsche Forschungsgemeinschaft*.

## REFERENCES

1. **Kauss, H.,** Biochemistry of osmotic regulation, in *Plant Biochemistry II*, Vol. 13, Northcote, D. H., Ed., University Park Press, Baltimore, 1977, 119.
2. **Kauss, H.,** Osmotic regulation in algae, in *Progress in Phytochemistry*, Vol. 5, Reinhold, L., Harborne, J. B., and Swain, T., Eds., Pergamon Press, New York, 1978, 1.
3. **Hellebust, J. A.,** Osmoregulation, *Annu. Rev. Plant Physiol.*, 27, 485, 1976.
4. **Schobert, B.,** Is there an osmotic regulatory mechanism in algae and higher plants?, *J. Theor. Biol.*, 68, 17, 1977.

5. **Zimmermann, U.**, Physics of turgor- and osmoregulation, *Annu. Rev. Plant Physiol.*, 29, 121, 1978.
6. **Gutknecht, J.**, Ion transport and turgor pressure regulation in giant algal cells, in *Membrane Transport in Biology*, Giebisch, G., Tosteson, D. C., and Ussing, H. H., Eds., Springer-Verlag, Berlin, 1978, 125.
7. **O'Donnell-Tormey, J. and Quigley, J. P.**, Detection and partial characterization of a chymostatin-sensitive endopeptidase in transformed fibroblasts, *Proc. Natl. Acad. Sci. U.S.A.*, 50, 344, 1983.
8. **Katunuma, N.**, New intracellular proteases and their role in intracellular enzyme degradation, *TIBS*, 2, 122, 1977.
9. **Kauss, H.**, Turnover of galactosyl-glycerol and osmotic balance in *Ochromonas*, *Plant Physiol.*, 52, 613, 1973.
10. **Kreuzer, H. P. and Kauss, H.**, Role of α-galactosidase in osmotic regulation of *Poterioochromonas malhamensis*, *Planta*, 147, 435, 1980.
11. **Dey, P. M.**, Involvement of D-galactokinase in the osmoregulation of *Poterioochromonas malhamensis*, *FEBS Lett.*, 112, 60, 1980.
12. **Quader, H. and Kauss, H.**, Die Rolle einiger Zwischenstoffe des Galaktosylglyzerinstoffwechsels bei der Osmoregulation in *Ochromonas malhamensis*, *Planta*, 124, 61, 1975.
13. **Spang, B., Claude, F., and Kauss, H.**, Partial purification and specificity of isofloridoside phosphatase, *Plant Physiol.*, 67, 190, 1981.
14. **Albrecht, G. J. and Kauss, H.**, Purification, crystallization and properties of a β-(1 → 3)-glucan phosphorylase from *Ochromonas malhamensis*, *Phytochemistry*, 10, 1393, 1971.
15. **Kauss, H., Thomson, K. S., Thomson, M., and Jeblick, W.**, Osmotic regulation: physiological significance of proteolytic and nonproteolytic activation of isofloridoside-phosphate synthase, *Plant Physiol.*, 63, 455, 1979.
16. **Kauss, H., Thomson, K. S., Tetour, M., and Jeblick, W.**, Proteolytic activation of a galactosyl transferase involved in osmotic regulation, *Plant Physiol.*, 61, 35, 1978.
17. **Kauss, H. and Quader, H.**, *In vitro* activation of a galactosyl transferase involved in the osmotic regulation of *Ochromonas*, *Plant Physiol.*, 58, 295, 1976.
18. **Kauss, H.**, Sensing of volume changes by *Poterioochroomonas* involves a $Ca^{2+}$-regulated system which controls activation of isofloridoside-phosphate synthase, *Plant Physiol.*, 68, 420, 1981.
19. **Thomson, K. S.**, Purification of UDPgalactose: *sn*-glycerol-3-phosphate α-D-galactosyltransferase from *Poterioochromonas malhamensis*, *Biochim. Biophys. Acta*, 759, 154, 1983.
20. **Köhle, D. and Kauss, H.**, Purification of a membrane-derived proteinase capable of activating a galactosyl-transferase involved in volume regulation, *Biochim. Biophys. Acta*, 799, 59, 1984.
21. **Kauss, H. and Thomson, K. S.**, Biochemistry of volume control in *Poterioochromonas*, in *Plasmalemma and Tonoplast: Their Functions in the Plant Cell*, Marmé, D., Marrè, E., and Hertel, R., Eds., Elsevier Biomedical Press, B. V., 1982, 255.
22. **Kauss, H.**, Volume regulation in *Poterioochromonas*. Involvement of calmodulin in the $Ca^{2+}$-stimulated activation of isofloridoside-phosphate synthase, *Plant Physiol.*, 71, 169, 1983.
23. **Kauss, H.**, Volume regulation: activation of a membrane associated cryptic enzyme system by detergent-like action of phenothiazine drugs, *Plant Sci. Lett.*, 26, 103, 1982.
24. **Mazzei, G. J., Katoh, N., and Kuo, J. F.**, Polymyxin B is a more selective inhibitor for phospholipid-sensitive $Ca^{2+}$-dependent protein kinase, *Biochem. Biophys. Res. Commun.*, 109, 1982, 1129.
25. **Kauss, H., Köhle, H., and Jeblick, W.**, Proteolytic activation and stimulation by $Ca^{2+}$ of glucan synthase from soybean cells, *FEBS Lett.*, 158, 84, 1983.
26. **Kauss, H. and Rausch, U.**, Compartmentation of $Ca^{2+}$ and its possible role in volume regulation of *Poterioochromonas*, in *Compartments in Algal Cells and their Interaction*, Wiessner, W., Robinson, D., and Starr, R. C., Eds., Springer-Verlag, Berlin, 1984, 149.
27. **Wise, B. C., Glass, D. B., Jen Chou, C. H., Raynor, R. L., Katoh, N., Schatzman, R. C., Turner, R. S., Kibler, R. F., and Kuo, J. F.**, Phospholipid-sensitive $Ca^{2+}$-dependent protein kinase from heart, *J. Biol. Chem.*, 257, 8489, 1981.
28. **Michell, B.**, $Ca^{2+}$ and protein kinase C: two synergistic cellular signals, *TIBS*, 8, 263, 1983.
29. Unpublished results.

Chapter 7

# PROTEOLYTIC ENZYMES OF LEGUME NODULES AND THEIR POSSIBLE ROLE DURING NODULE SENESCENCE

## C. P. Vance, P. H. Reibach, and W. R. Ellis

## TABLE OF CONTENTS

| | | |
|---|---|---:|
| I. | Introduction: Legume Nodule Ontogeny and Senescence | 104 |
| II. | Proteases and Senescence of Spherical Nodules | 112 |
| III. | Proteases and Senescence of Elongate Nodules | 118 |
| IV. | Conclusions | 122 |
| | References | 122 |

## I. INTRODUCTION: LEGUME NODULE ONTOGENY AND SENESCENCE

Proteolytic enzymes are important in plant development and senescence.[1,2] Their occurrence has been documented in roots, leaves, flowers, and fruits of numerous species. The role of proteolytic enzymes in plant metabolism has received extensive interest in the last 20 years,[1] particularly with respect to mobilization of protein reserves in seeds during germination (see Chapters 1 and 2) and mobilization of leaf protein (see Chapter 3) during grain filling of annual crops.[1,2] However, to our knowledge, only six articles have been published that directly deal with proteolytic enzymes in legume root nodules.[3-8]

Rhizobium-legume-induced root nodules are highly organized hyperplastic, hypertrophic tissue masses derived from root cortical cells.[9-11] Their development is dependent upon interactions between the plant and the bacterium. The root nodule provides an ecological niche for the bacterium, while the bacterium provides the plant with a source of fixed nitrogen. Nodules are divided into three major groupings based upon shape and meristematic activity: (1) elongate-cylindrical nodules with indeterminate apical meristematic development, such as alfalfa, pea, and clover; (2) spherical nodules with transient internal meristematic development such as trefoil, common bean, and soybean; and (3) collar-type nodules that encircle the tap root, such as lupin.[9-11] Several studies have described the ontogeny and senescence of both elongate-cylindrical nodules and spherical nodules.[12-18] The ontogeny of collar-type nodules is less well understood;[11] therefore, our discussion of nodule proteolytic enzymes centers on elongate-cylindrical and spherical nodules. Before considering nodule proteolytic enzymes, some concept of nodule structure during ontogeny is needed.

Elongate cylindrical nodules are characterized by two colored regions: a white region, which includes the nodule cortex, meristem, and zone of infection thread invasion; and a pink region with cells containing leghemoglobin and bacteroids in various stages of development (Figure 1). As nodules senesce, a third region, green in color, forms at the proximal end of the nodules and contains senescent cells with disintegrating bacteroids (Figures 1 and 2). The nodule meristem develops and proliferates in advance of the infection threads. Complete senescence of the nodule occurs when the nodule meristem becomes nonfunctional. Nodule cells proximal to and contiguous with the meristem are invaded by infection threads containing rhizobia. In this thread invasion zone, rhizobia are released from infection threads into host cell cytoplasm. Each bacterium is tightly enclosed by the peribacteroid membrane. As nodules increase in size and progress from early to late symbiotic development, cells containing bacteroids increase several-fold in volume, and bacteroids increase markedly in size and fill the infected cells.

Spherical nodules are composed of an external cortical cell layer in which vascular bundles are arranged around a central zone comprised of both infected and uninfected cells (Figure 3). In contrast to elongate-cylindrical nodules, spherical nodules do not contain various developmental zones within a single nodule. Small, newly formed nodules are white and contain numerous actively dividing cells, prolific infection thread development, and a cortex. Meristematic activity occurs in all planes and is limited to the initial stages of nodule development. As nodule development proceeds, bacteria are released and increase in number, and nodules appear pink because of leghemoglobin. Senescent nodules are characterized by a green or dark brown color, disintegration of nodule tissue, and in some instances lysis of bacteroids (Figure 4). Nodules at all stages of development can be found on a single mature plant.

Although the exact sequence of events during nodule senescence remains unclear, the end result in both elongate-cylindrical and spherical nodules is similar. Host cell cytoplasm appears less dense and numerous vesicles form. Cellular organelles are conspicuously absent. Frequent interruptions in the tonoplast and peribacteroid membranes are evident. Eventually, senescent cells may contain convoluted membrane fragments with a few deteriorating bacteroids and fragments of cytoplasm. Bacteroids in cylindrical-elongate nodules undergo lysis. However, the fate of bacteroids in spherical nodules is less clear. Bacteroids in the spherical nodules of birdsfood trefoil undergo lysis.[18] In contrast, soybean nodule bacteroids do not decrease in protein and may revert to the free-living form.[7,8] Infection threads and endophytic bacteria appear to remain intact in senescent cells of both nodule types.

Although there are few studies that directly address proteolytic enzymes of legume root nodules, the effect of defoliation, shading, applied nitrogen, diseases, and reduced effectiveness on nodule function and structure have been carefully studied. Shading and defoliation reduce leghemoglobin content and induce nodule senescence in both spherical and elongate nodules.[3,7,18-26] Vance et al.[3] demonstrated that defoliation induced nodule senescence and leghemoglobin destruction; as shoots regrew, nodule senescence was alleviated and leghemoglobin content recovered. Nodules of soybean plants lost soluble protein and nitrogenase activity when shoots were placed in darkness.[7] When plants were returned to control irradiances, nodules recovered both soluble protein and nitrogenase activity. Arturri et al.[22] reported that soybean nodules turn brown and/or green when shoots were kept in darkness. They also reported a decrease in leghemoglobin. Pea plants had green nodules with reduced bacteroid numbers and leghemoglobin after placing shoots in darkness for 3 days.[21] Defoliation of birdsfoot trefoil caused a reduction in nitrogenase activity, nodule soluble protein, and nodule fresh weight.[18] The percentage of senescent green nodules increased upon defoliation, then decreased as new nodules formed during shoot regrowth.

Nodule function is reduced and structure is impaired when nitrate and ammonium are applied to active nitrogen-fixing legumes.[27-33] Nodule meristem cytoplasm was disorganized in *Medicago tribuloides* and *Trifolium subterraneum* after treatment with ammonium nitrate.[27] Bacteroid enlargement and peribacteroid membrane formation were blocked. Host cell cytoplasm and organelles rapidly deteriorated in the presence of ammonium nitrate. Pea root nodule leghemoglobin content and nitrogenase activity were significantly reduced within 4 days of exposure to either ammonium or nitrate.[29] Applied nitrate caused a rapid decline in nitrogen fixation potential of alfalfa.[32] Decreases in alfalfa nodule nitrogen fixation capacity were accompanied by decreases in nodule soluble protein, glutamine synthetase, and glutamate synthase. As nitrate concentrations were reduced in the rooting media, nodule function recovered. Streeter[33] demonstrated that nitrate suppressed nodule weight in soybeans.

The genotype of the host or *Rhizobium*, plant diseases, and water and temperature stresses also cause alterations in nodule soluble protein and nodule structure. Ineffective nodules, whether induced by the host plant genome, bacterial genome, or the environment, are accompanied by premature senescence of nodule tissue, and reduced protein concentrations compared to effective nodules.[34-37] Nodule aging is accompanied by reduced leghemoglobin, protein loss, and lysis of bacteroids.[3,5,8,38,43] Plant diseases caused by viruses, nematodes, and fungi induce premature senescence, reduced protein, and legume nodule dysfunction.[43-48] Water and temperature stress also cause nodule senescence, bacteroid lysis, and losses in nodule protein.[43,49-52]

Unfortunately, the role of proteolytic enzymes in nodule development, function, and senescence has been evaluated in only a few of the studies cited above. In the sections that follow, we will review studies of proteolytic enzymes in both elongate-cylindrical and spherical nodules. We will attempt to show where gaps in our understanding occur and thereby stimulate interest in further research.

## II. PROTEASES AND SENESCENCE OF SPHERICAL NODULES

Senescence of spherical nodules can occur concomitantly with whole plant senescence at the end of the growing season. Spherical nodules also senesce in response to various environmental treatments such as detopping, shading, and combined nitrogen as reviewed

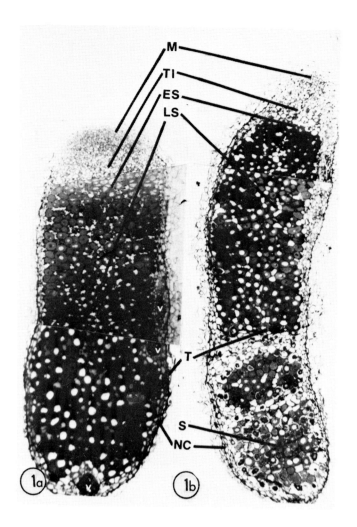

FIGURE 1. Light micrograph montage of median longitudinal sections of alfalfa nodules collected from plants subjected to shoot removal. Zones illustrated are: meristem (M); thread invasion (TI); early symbiotic (ES); late symbiotic (LS); transition (T); and senescent zone (SZ). (Magnification × 64). (1a) Nodule collected 1 day after the shoot was removed. Bacteroid containing cells in the late symbiotic stage of development are found throughout the central mass of the nodule. A small transition zone and senescent zone are seen at the base of the nodule. (1b) Nodule collected 4 days after a foliage harvest. Nodules increased in length; however, fewer bacteroid-containing cells were found in the central tissue masses of nodules; transition zones advanced distally. Senescent bacteroid-containing cells are present in these large senescent zones. (1c) Nodule collected 10 days after foliage harvest. Nodules increased in length, and fewer mature bacteroid-containing cells were found in the late symbiotic zones. Senescent bacteroid-containing cells fill approximately 75% of the central mass of this nodule. (1d) Nodule collected 18 days after foliage harvest. Nodules increased in length, and mature bacteroid-containing cells in late symbiotic stage increased as the rate of senescence slowed. Transition and senescent zone occupied areas equal to those observed in day 10 nodules. Senescent bacteroid-containing cells have collapsed.

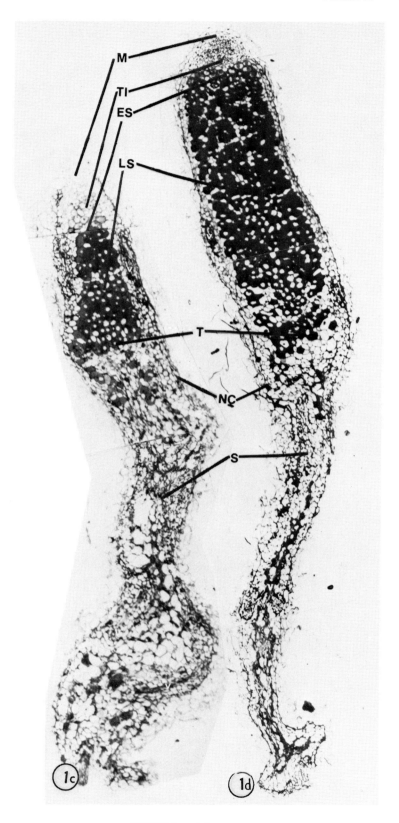

FIGURE 1c and d

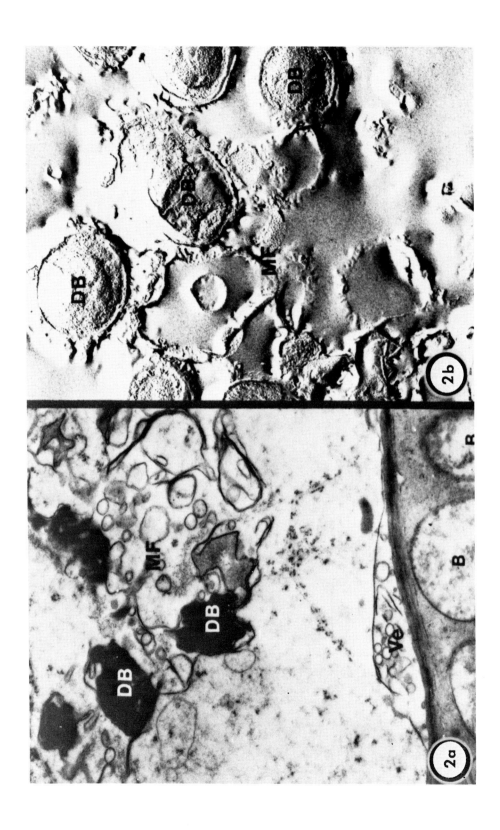

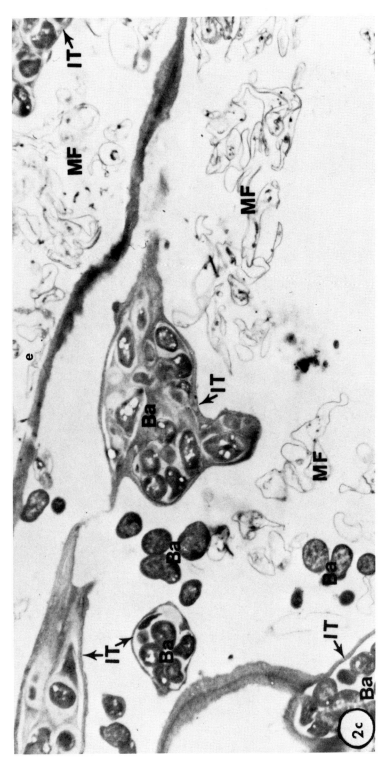

FIGURE 2. Senescence in nodule bacteroid-containing cells. (2A) Bacteroid-containing cells from the transition zone of a nodule from an unharvested control plant. A normal symbiotic cell containing bacteroids (B) lies adjacent to a deteriorating cell. Numerous vesicles (Ve) are present in the deteriorating cell. Cytoplasm has almost completely deteriorated leaving membrane fragments (MF). Deteriorated bacteroids (DB) appear dense as nucleoid material coalesces. (Magnification × 20,000). (2B) Freeze-fracture replica of a bacteroid-containing cell from the senescent zone of a nodule from harvested plants collected on day 10 (Figure 1c). The cell has little if any cytoplasm. Bacteroids (B) are reduced in size and membrane fragments are scattered throughout the cell. (Magnification × 24,000). (2c) Cells in the senescent zone of a nodule from harvested plants collected on day 18 (Figure 1d). No bacteroids can be seen in the cells and cytoplasm is absent. Membrane fragments (MF) are scattered throughout the cell. Apparently undamaged infection threads (IT) are releasing endophytic bacteria (Ba) into senescent host cells. (Magnification × 8,000).

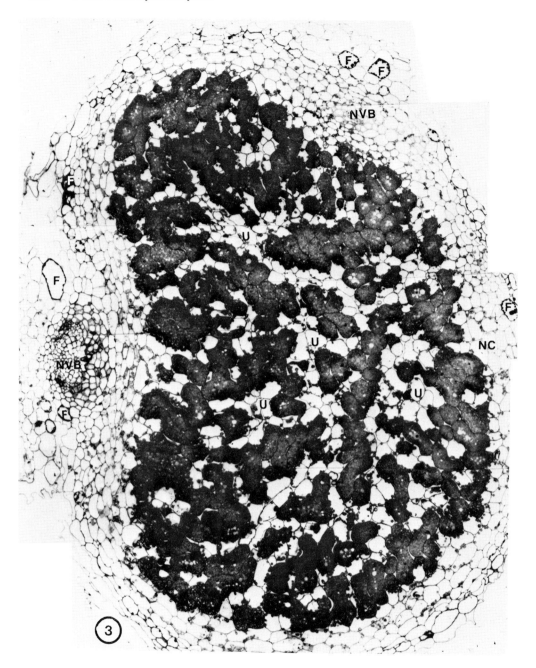

FIGURE 3. Light micrograph montage of a median longitudinal section of a large pink birdsfoot trefoil nodule. Nodule is comprised of a central zone with infected cells containing bacteroids, uninfected cells (U) containing starch (S) granules, and a nodule cortex (NC). Nodule vascular bundles (NVB) are embedded in the cortex and some cortical cells contain flavolans (F). (Magnification × 140.)

before. Both types of senescence are characterized by reductions in nitrogenase activity, respiration, and nodule growth. These characteristics are generally easily observable and have been well-documented. These morphological and biochemical changes are similar to those observed during senescence of other determinate plant organs.[53]

Changes in levels of nucleic acids, proteins, and intermediary metabolites that are observed during the senescence of other plant organs have not been extensively studied during spherical nodule senescence. Studies of nodule senescence are further complicated by the presence of *Rhizobium* in the nodules, which do not necessarily senesce, but may continue to grow in a saprophytic fashion. For example, bacteroid nucleic acid levels during spherical nodules senescence did not decline, while nucleic acid levels in bacteria/bacteroids from elongate nodules did decline.[19]

Proteolysis and its regulation is a complicated phenomenon involving events separated spatially and temporally. Protease activity has been monitored throughout nodule development and senescence, but little work has been reported on the subcellular localization of this activity. Few studies have dealt specifically with protease activity in legume nodules. However, there are numerous reports of declines in total nodule protein, nodule leghemoglobin content, and activity of nodule enzymes involved in intermediary metabolism. These declines in protein content can be due to degradation by proteolysis and/or declines in protein synthesis.

Pfeiffer et al.[8] investigated proteolytic activity of soybean nodules during senescence. Soybean plants were grown and allowed to senesce under natural conditions. Nodule characteristics that were monitored included acetylene reduction activity, number and fresh weight of nodules, soluble protein content, leghemoglobin content, and protease activities of the cytosol and bacteroid fractions (Table 1). Total soluble protein and leghemoglobin declined significantly during the latter stages of plant maturity. This decline was most dramatic following physiological maturity of the plant (pods and leaves yellowing). Azocasein protease specific activity, expressed as units per milligram protein, increased at physiological maturity; however, if expressed as units per gram fresh weight nodules, the level of protease activity remained relatively constant. Little change was found for bacteroid protein or protease levels, suggesting that bacteroids do not disintegrate during nodule senescence (Table 2).

Proteases were also characterized with respect to substrate specificity.[8] Activity was found with L-leu-βNA and Bz-L-Arg-βNA as substrates, but the highest activity was obtained with hemoglobin. Low activity was found with azocasein as the substrate. Nodule autolytic activity was lower than hemoglobin degradation, but significantly higher than either azocasein degradation or the degradation of synthetic substrates.

Similar developmental profiles were reported for dark-induced nodule senescence of soybean[7] (Figure 5). Again, large increases in protease specific activity were observed during senescence, but total protease activity changed little throughout the dark treatment and during rejuvenation of nodule activity by returning plants to the light. Interestingly, all nodule functions that declined during the dark-induced senescence returned to normal, pre-dark levels. This would suggest that depletion of photosynthate reduced nodule functions to counteract the proteolytic degradation, which occurs during normal nodule ontogeny.

In cowpea nodules, autodigestion and hemoglobin hydrolytic activity of nodule fractions rose to a maximum at flowering and fell to a low level during root senescence (Figure 6).[6] Hemoglobin hydrolysis by root extracts increased until late in senescence of the root and nodule tissue. Hydrolytic activity in nodule cortex and cytosol fractions followed the same trend as total nodule soluble protein. Bacteroid proteolytic activity was much lower than in the other fractions, but generally followed the same pattern of development as the other fractions, suggesting again that proteolytic enzymes are present throughout the growth cycle of spherical nodules, and that reduction in nodule protein concentration may occur because of decreased nodule capability to replenish degraded protein.

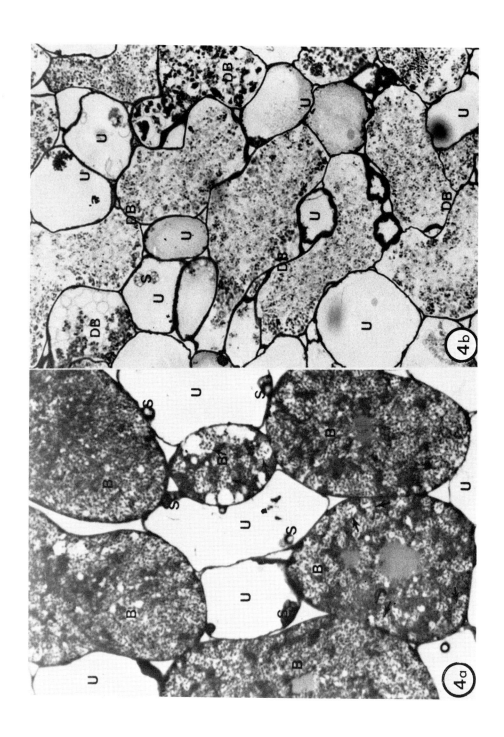

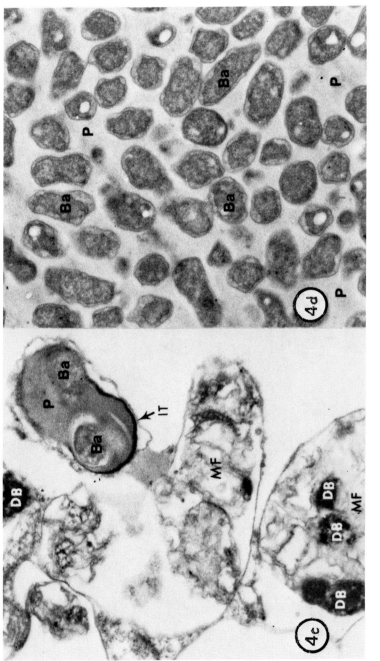

FIGURE 4. Senescence in spherical nodules of birdsfoot trefoil. (4A) During early stages of senescence in cells containing bacteroids (B) large vacuoles (arrows) are formed by aggregation and vesiculation of peribacteroid membrane. Uninfected cells (U) containing starch (S) have little cytoplasm evident. (Magnification × 700). (4B) Bacteroids in the center of senescent green nodules aggregate and lyse. Cells appear misshapen and crushed in these advanced stages of senescence. (Magnification × 700). (4c) Infected cell from a senescent green nodule. No intact bacteroids (DB, disintegrating bacteroids) can be seen in cells and cytoplasm is absent. Membrane fragments (MF) are scattered throughout the cell. Apparently undamaged infection threads (IT) containing bacteria (Ba) can be found in senescent cells. (Magnification × 20,800). (4d) Some central zone cells of senescent green nodules contain saprophytic bacteria (Ba) surrounded by a polysaccharide matrix (P). (Magnification × 16,900).

## Table 1
### TOTAL FRESH WEIGHT, NUMBER OF SOYBEAN ROOT NODULES, AND ENDOPEPTIDASE ACTIVITIES OF BACTEROID-FREE NODULE CYTOSOL FROM PLANTS HARVESTED FROM EARLY VEGETATIVE GROWTH THROUGH HARVEST MATURITY

| Physiological stage[a] | Nodule fresh wt[b] (g) | Nodule[b] no. | Nodule dry wt[b] (%) | mg/g Fresh wt of nodule | | | | Units/mg protein[b] | |
|---|---|---|---|---|---|---|---|---|---|
| | | | | Soluble protein[b] | Leghemoglobin[b] | Leghemoglobin (% soluble protein)[b] | Azocasein[b] | Bz-L-Arg[c] βNA (pH 9.8) | Bz-L-Arg[c] βNA (pH 7.5) |
| Vegetative | 2.3 C | 164 B | — | 7.53 AB | 2.95 AB | 39.2 BCD | 0.23 C | 0.54 ABC | 0.28 BC |
| Full flower | 5.9 C | 312 B | 17.4 A | 7.62 A | 3.03 A | 39.8 AB | 0.25 C | 0.63 A | 0.33 B |
| Early pod | 13.8 BC | 471 AB | 17.1 A | 5.26 BC | 2.54 ABC | 48.3 A | 0.43 AB | 0.73 A | 0.43 A |
| Full pod | 19.0 AB | 609 A | 18.3 A | 6.51 AB | 2.29 BC | 35.2 BC | 0.35 B | 0.59 AB | 0.30 BC |
| Early seed | 18.7 AB | 583 A | 18.4 A | 6.82 AB | 2.02 C | 29.6 CDE | 0.33 BC | 0.51 BC | 0.29 BC |
| Full seed | 19.5 AB | 656 A | 18.3 A | 7.10 AB | 1.99 C | 28.0 DEF | 0.33 BC | 0.45 C | 0.24 C |
| Physiological maturity | 26.1 A | 712 A | 16.7 A | 4.71 C | 1.24 D | 26.3 F | 0.51 A | 0.46 C | 0.30 BC |
| Harvest maturity | 9.1 C | 315 B | 15.9 A | 4.68 C | 1.30 D | 27.8 EF | 0.54 A | 0.46 BC | 0.29 BC |

[a] Plants were grown in an environmentally controlled growth chamber at 26 ± 2°C on a 14 hr photoperiod.
[b] Data were analyzed using Duncan's multiple range test. Means denoted by the same letter are not significantly different at the 5% level of probability.
[c] One unit of activity against azocasein is expressed as the amount of enzyme which produced a △A440 of 1.0/hr, while one unit of activity against Bz-L-Arg-βNA is expressed as the amount of enzyme required to liberate 1 µmol product per hour.

Data from Pfeiffer, N. E., Torres, C. M., and Wagner, F. W., *Plant Physiol.*, 71, 797, 1983. With permission.

## Table 2
### SOLUBLE PROTEIN CONTENT AND PROTEINASE ACTIVITY OF BACTEROIDS EXTRACTS FROM SOYBEAN NODULES HARVESTED THROUGHOUT THE GROWING SEASON

| Physiological stage | Soluble protein (mg/g fresh wt of nodule) | Protein-digesting activity[a] (units × $10^2$/mg protein) | | |
|---|---|---|---|---|
| | | Hemoglobin[b] | Autolysis[b] | Azocasein[c] |
| Vegetative | 5.14 A | 11.08 A | 5.17 AB | 1.63 AB |
| Full flower | 5.78 A | 8.39 A | 5.66 A | 1.49 AB |
| Early pod | 6.08 A | 5.99 A | 6.89 A | 1.45 B |
| Full pod | 5.84 A | 7.64 A | 7.19 A | 1.60 AB |
| Early seed | 6.06 A | 7.30 A | 6.56 A | 1.46 B |
| Full seed | 5.78 A | 8.00 A | 6.71 A | 1.62 AB |
| Physiological maturity | 6.10 A | 7.76 A | 7.76 A | 1.75 A |
| Harvest maturity | 6.06 A | 11.73 A | 1.35 A | 1.70 AB |

[a] Data were analyzed using Duncan's multiple range test. Means denoted by the same letter were not significantly different at the 5% level of probability.
[b] One unit of hemoglobin digesting or autolytic activity is defined as the amount of enzyme that released 1 μmol amino-N/hr.
[c] One unit of activity against azocasein is expressed as the amount of enzyme which produced $\triangle$A440 of 1.0/hr.

Data from Pfeiffer, N. E., Torres, C. M., and Wagner, F. W., *Plant Physiol.*, 71, 797, 1983. With permission.

Unpublished results from our laboratory with adzuki bean nodules showed a reduction in total protein at 9 weeks after planting (Figure 7). This decline in total nodule protein was accompanied by an increase in azocasein protease specific activity and total activity. No characterization of the proteases were performed. While these data are not in complete agreement with data from soybean and cowpea, all of these studies demonstrate that in vitro protease activity is high during nodule senescence.

Proteases from soybean and french bean nodules have been characterized by Malik et al.[4] (Tables 3 and 4) and by Pladys and Rigaud.[5] Soybean nodules were found to contain endopeptidase, aminopeptidase, and carboxypeptidase activities. Endopeptidase activity towards azocasein and benzoyl-Arg-βNA were detected. The authors proposed that the low activity observed towards azocasein could be due to the number of susceptible bonds, but multiple endopeptidases could not be ruled out. Aminopeptidase activity towards L-Leu-, L-Ala-, or L-Val-βNA was detected. Highest activity was for L-Leu-βNA while very low activity was observed for L-Val-βNA. Carboxypeptidase activity was difficult to quantitate due to the presence of interfering substances in the nodule extracts. Electrophoresis of crude cytosol extracts at pH 8.8 resolved three zones of endopeptidase activity (Figure 8). None of the protease bands that were stained *in situ* were able to hydrolyze all of the substrates tested.

Gels were also stained for aminopeptidase activity. Four bands of activity were observed, none of which had an acid pH optimum (Table 3). All of the endopeptidases were classified as serine proteases, while two of the aminopeptidases were sensitive to sulfhydryl-group inhibitors and had a metal requirement. While no correlations between specific proteases

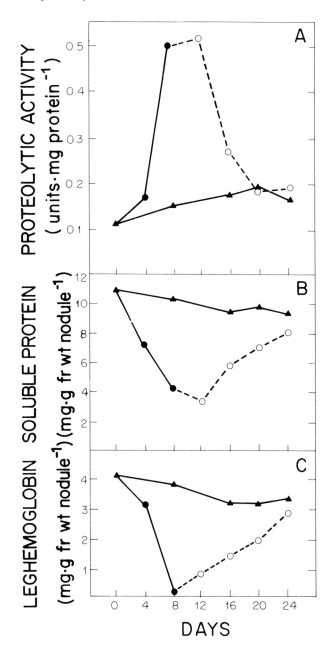

FIGURE 5. Endopeptidase activity of soybean nodule extracts using azocasein as substrate (A), total soluble protein in nodule cytosol determined by the dye binding assay (B), and leghemoglobin concentration determined by pyridine hemochrome method (C). Symbols are (▲) nodules maintained under a normal 14-hr photoperiod; (●) nodules from plants kept in total darkness; and (○) nodules from plants kept in darkness for 8 days, then allowed to recover under normal 14-hr photoperiod for up to 16 days. (From Pfeiffer, N. E., Torres, C. M., and Wagner, F. W., *Plant Physiol.*, 71, 393, 1983. With permission.)

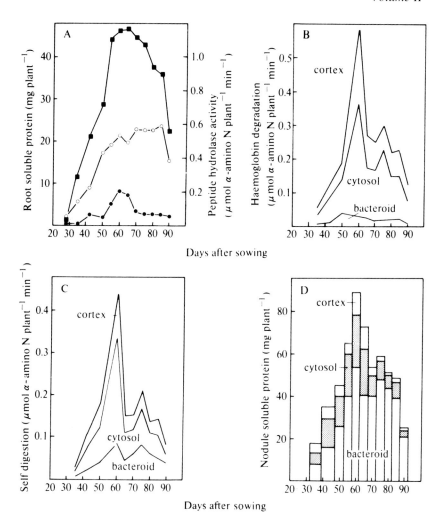

FIGURE 6. Changes with plant age in peptide hydrolase activities and soluble protein levels of extracts from roots and nodules of nodulated plants of cowpea (Vita 3). (A) Autodigestion (closed circles), hemoglobin degradation (open circles), and soluble protein (solid squares) of root extracts. Hemoglobin degradation (B), autodigestion (C), and soluble protein (D) of bacteroid, cytosol, and cortex fraction of nodules (data plotted cumulatively). (From Peoples, M. B., Pate, J. S., and Atkins, C. A., *J. Exp. Bot.*, 34, 563, 1983. With permission.)

and stage of nodule development were performed, the methodologies developed should prove useful for subsequent research on the appearance of proteases specific to nodule senescence.

Pladys and Rigaud[5] purified proteases 50-fold from *Phaseolus vulgaris* nodules. The pH optimum with leghemoglobin as substrate was 3.5 to 4.0, with similar results obtained for hemoglobin digestion. Inhibition studies suggested that the purified protein was a serine protease. The Km for leghemoglobin was estimated to be 0.60 m$M$, which was similar to hemoglobin. Differences between pH optima for proteases from this study and that of Malik et al.[4] could be due to the different growth stage of the tissue.

Due to the lack of information, the precise role of proteases in spherical nodule senescence is not known. Protein degradation occurs during senescence. It is not known whether this degradation is due to synthesis of specific proteases or the lack of ability to replenish degraded proteins by protein synthesis. The release of proteases from subcellular components cannot be ruled out. Based on current data, the involvement of proteases in nodule senescence

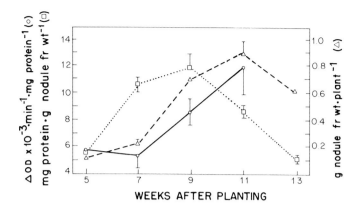

FIGURE 7. Adzuki bean nodule azocasein protease activity (○), nodule soluble protein (□), and nodule fresh weight (△) throughout the growing season.

Table 3
SUMMARY OF pH OPTIMA OF NODULE PEPTIDASES

| Enzyme fraction | pH Optimum | pH Range ($\pm 50\%$ optimum) |
|---|---|---|
| EP-Arg-1 | 9.3 | 8.2—10.2 |
| EP-Arg-2 | 7.5, 9.8 | 6.3—10.9 |
| AP-Leu-1 | 6.7, 8.0 | 5.5—9.3 |
| AP-Leu-2 | 6.7 | 5.5—7.4 |
| AP-Ala | 9.1 | 6.6—10.0 |
| AP-Val | 8.0 | 7.0—9.0 |

*Note:* Enzymes were separated by slab gel electrophoresis. Bands of peptidohydrolase activity were extracted with 25 m$M$ Na-phosphate (pH 7.0), then assayed using the appropriate substrate in 100 m$M$ buffer of the desired pH. Buffer salts used were citrate, $pka_2 = 4.74$, $pka_3 = 5.40$; phosphate, $pka_2 = 7.21$; Tris, $pka = 8.1$; borate, $pka = 9.24$.

Data from Malik, N. S. A., Pfeiffer, N. E., Williams, D. R., and Wagner, F. W., *Plant Physiol.*, 68, 386, 1981. With permission.

appears to be the degradation of nodule proteins during a stage of development when protein synthesis is not capable of regenerating functional proteins. The effect of applied nitrogen on soybean nodule protein synthesis supports this interpretation, since both $NH_4^+$ and $NO_3^-$ cause significant decreases in soybean nodule protein synthesis as measured by incorporation of $^{35}S$.[54]

## III. PROTEASES AND SENESCENCE OF ELONGATE NODULES

Senescence and protein degradation in spherical nodules generally occur during whole plant senescence. However, elongate nodules, particularly those of perennial legumes, may undergo sequential senescence and recovery several times during the life span of the plant.[10,11]

## Table 4
## EFFECT OF MODIFYING AGENTS ON PEPTIDASE ACTIVITIES PURIFIED FROM SOYBEAN NODULE EXTRACTS

| | | Percent remaining activity | | | | | | |
|---|---|---|---|---|---|---|---|---|
| Modifying agent | Conc. mM | EP-Arg-1 | EP-Arg-2 (pH 7.5) | EP-Arg-2 (pH 9.8) | AP-Leu-1 (pH 6.7) | AP-Leu-1 (pH 8.0) | AP-Leu-2 | AP-Ala | AP-Val |
| DFP[a] | 1 | 0[c] | 0 | 0 | 82[d] | 75 | 80 | 98 | 75 |
| PCMB[b] | 0.025 | 89 | 66 | 73 | 0 | 17 | 287 | 64 | 0 |
| Dithioerythritol | 1.25 | 120 | 102 | 97 | 223 | 140 | 111 | 108 | 139 |
| | 10.0 | 60 | 84 | 84 | 24 | 48 | 78 | 10 | 1 |
| 1,10-Phenanthroline | 1.0 | 77 | 99 | 93 | 23 | 25 | 87 | 157 | 36 |

[a] Diisopropylphosphorofluoridate was preincubated for 30 min at room temperature with 20 mM cysteine. Control assay contained 20 mM cysteine.
[b] p-Chloromercuribenzoate assays were performed with Bz-L-Arg-βNA for endopeptidases.
[c] Standard deviations were less than 6% of the mean for endopeptidases.
[d] Standard deviations were less than 8% of the mean for aminopeptidases.

Data from Malik, N. S. A., Pfeiffer, N. E., Williams, D. R., and Wagner, F. W., *Plant Physiol.*, 68, 386, 1981. With permission.

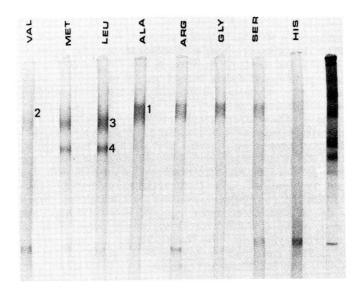

FIGURE 8. Disc gel electrophoresis of soybean nodule cytosol extract. Gels were incubated with various L-aminoacyl-βNAs (0.4 mg/mℓ H$_2$O) for 15 min. The gel labeled Val was incubated with L-Val-βNA for 1 hr at ambient temperature. After incubation, the substrate gels were allowed to react with 0.1% Fast Garnet GBC salt for 10 min at ambient temperature. The identity of the amino acid is given above each gel. The gel at the extreme right of the photo was stained with Coomassie blue (0.02% in H$_2$O). Numbers in the photograph indicate: (1) AP-Ala, (2) AP-Val, (3) AP-Leu-1, and (4) AP-Leu-2. (From Malik, N. S. A., Pfeiffer, N. E., Williams, D. R., and Wagner, F. W., *Plant Physiol.*, 68, 386, 1981. With permission.)

This unique adaptive feature results from the indeterminant nature of the apical meristem of elongate nodules (Figure 1).

The data base for proteolytic enzymes in elongate nodules is even smaller than that for spherical nodules. Although several studies have evaluated morphology, leghemoglobin content, and nitrogenase function during the ontogeny of elongate nodules,[3,12-14,20,27,32,39] only one study, to our knowledge, has addressed proteolytic enzyme activity in elongate nodules.

Vance et al.[3] followed alfalfa nodule structure and function after shoot removal and through the regrowth cycle. Nodules begin to show both structural and physiological signs of senescence shortly after shoot removal. Both specific and total azocasein proteolytic activity in nodules increased rapidly from day 1 after shoot removal to a maximum at day 7, then declined rapidly until day 13 and stabilized at a value slightly greater than that of control nodules (Figure 9). Protease activity in control nodules remained unchanged. Nodule soluble protein in plants with shoots harvested inversely reflected nodule protease activity and had a pattern similar to that of nodule leghemoglobin concentration (Figures 10 and 11). Soluble protein in nodules of harvested plants declined rapidly between days 1 and 7, more slowly between days 7 and 10, and then increased rapidly from day 10 to 22. Soluble protein in nodules of control plants varied little during the experiment.

These changes in nodule protease and soluble protein were reflected in nodule structure and development (Figure 1). Initial stages of nodule senescence were apparent 4 days after shoot removal and by day 10 only a small number of nodule cells immediately adjacent to the nodule meristem contained bacteroids. Eighteen days after shoot removal, coinciding with increased shoot regrowth, signs of nodule regeneration were evident. The close asso-

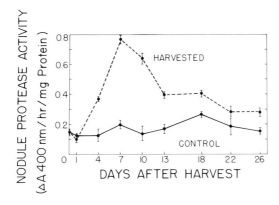

FIGURE 9. Effect of shoot harvest on alfalfa nodule azocasein protease activity. Each point is the mean ±SE of three replicates.

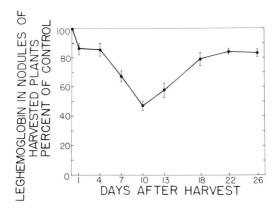

FIGURE 10. Effect of shoot harvest on leghemoglobin concentration in alfalfa nodules. Each point is the mean ±SE of three replicates.

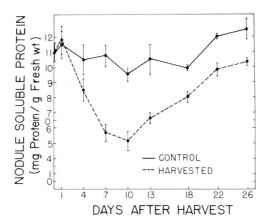

FIGURE 11. Patterns of alteration of alfalfa nodule soluble protein as influenced by shoot harvest. Each point is the mean ±SE of three replicates.

ciation between the physiological signs of senescence, structural deterioration and recovery, and the pattern of nodule proteolysis in alfalfa nodules indicates that proteases are involved in degradation of symbiotic tissue. Increased vesiculation (Figure 2) in senescent cells suggests that proteases may be produced and released from subcellular compartments in response to factors initiating senescence. The close association between shoot growth and development and nodule proteolysis suggests a role for current photosynthate in maintaining nodule integrity. However, a crucial role for hormones is not negated by this observation.

Bacteroid senescence appears to accompany nodule proteolysis and deterioration in elongate nodules. Leghemoglobin decreased within 2 days of placing pea plants in complete darkness.[21] Nodule leghemoglobin decreases were accompanied by a decrease in bacteroid number and an increase in the number of saprophytic bacteria in the nodule. The alterations in pea nodule leghemoglobin and bacteroid number upon dark-induced senescence were also accompanied by decreases in soluble asparagine, homoserine, γ-aminobutyric acid, and ethanolamine. Vance et al.[12] showed that plant cytoplasm and bacteroids were degraded in senescent nodule cells of alfalfa leaving only remnants of peribacteroid membranes (Figure 2). Dark-induced senescence decreased alfalfa nodule bacteroid size and nucleic acid content,[19,20] supporting the concept that senescence of elongate nodules is accompanied by bacteroid degradation. To date the proteolytic and/or hydrolytic enzymes responsible for bacteroid deterioration have not been evaluated.

## IV. CONCLUSIONS

After examining the limited data base, we think it prudent to pursue several fields of investigation to assess the significance of nodule proteases to nodule ontogeny and development. These include (1) obtaining general information about numbers and types of proteases in both spherical and elongate nodules; (2) studying developmental profiles of proteases in both elongate and spherical nodules; (3) determining whether nodule proteases are activated or synthesized *de novo* during nodule ontogeny; (4) evaluating the occurrence of protease inhibitors throughout nodule development; (5) determining the subcellular compartmentalization of nodule proteases; (6) separating host plant and bacterial components of nodule proteolysis; and (7) comparing and contrasting proteolytic enzymes between effective and ineffective nodules controlled by both host and *Rhizobium*.

It is clear that proteolysis and proteolytic enzyme activity in legume nodules is an emerging area. Conflicting evidence and the small quantity of available information clearly indicate the need for additional efforts to determine whether the senescence of nodules during legume ontogeny is a major limitation to nitrogen fixation.

## REFERENCES

1. **Ryan, C. A.**, Proteolytic enzymes and their inhibitors in plants, *Annu. Rev. Plant Physiol.*, 24, 173, 1973.
2. **Ashton, F. M.**, Mobilization of storage proteins of seeds, *Annu. Rev. Plant Physiol.*, 27, 95, 1976.
3. **Vance, C. P., Heichel, G. H., Barnes, D. K., Bryan, J. W., and Johnson, L. E. B.**, Nitrogen fixation, nodule development and vegetative regrowth of alfalfa (*Medicago sativa* L.) following harvest, *Plant Physiol.*, 64, 1, 1979.
4. **Malik, N. S. A., Pfeiffer, N. E., Williams, D. R., and Wagner, F. W.**, Peptidohydrolases of soybean root nodules: identification, separation and partial characterization of enzymes from bacteroid free extracts, *Plant Physiol.*, 68, 386, 1981.
5. **Pladys, D. and Rigaud, J.**, Acid protease and leghemoglobin digestion in french bean nodules, *Z. Pflanzenphysiol.*, 108, 163, 1982.
6. **Peoples, M. B., Pate, J. S., and Atkins, C. A.**, Mobilization of nitrogen in fruiting plants of a cultivar of cowpea, *J. Exp. Bot.*, 34, 563, 1983.

7. **Pfeiffer, N. E., Malik, N. S. A., and Wagner, F. W.,** Reversible dark-induced senescence of soybean root nodules, *Plant Physiol.,* 71, 393, 1983.
8. **Pfeiffer, N. E., Torres, C. M., and Wagner, F. W.,** Proteolytic activity in soybean root nodules: activity in host cell cytosol and bacteroids throughout physiological development and senescence, *Plant Physiol.,* 71, 797, 1983.
9. **Newcomb, W.,** Nodule morphogenesis and differentiation, in *International Review of Cytology, Supplement 13, Biology of the Rhizobiaceae,* Giles, K. L. and Atherly, A. G., Eds., Academic Press, New York, 1981, 247.
10. **Vance, C. P.,** *Rhizobium* infection and nodulation: a beneficial plant disease?, *Annu. Rev. Microbiol.,* 37, 399, 1983.
11. **Dart, P.,** Infection and development of leguminous nodules, in *A Treatise on Dinitrogen Fixation, Section III: Biology,* Hardy, R. W. F. and Silver, W. S., Eds., John Wiley & Sons, New York, 1977, chap. 8.
12. **Vance, C. P., Johnson, L. E. B., Halvorsen, A. M., Heichel, G. H., and Barnes, D. K.,** Histological and ultrastructural observations of *Medicago sativa* root nodule senescence after foliage removal, *Can. J. Bot.,* 58, 295, 1980.
13. **Newcomb, W.,** A correlated light and electron microscopic study of symbiotic growth and differentiation in *Pisum sativum* root nodules, *Can. J. Bot.,* 54, 2163, 1976.
14. **Paau, A. S., Block, C. B., and Brill, W. J.,** Developmental fate of *Rhizobium meliloti* bacteroids in alfalfa nodules, *J. Bacteriol.,* 143, 1480, 1980.
15. **Goodchild, D. J. and Bergersen, F. J.,** Electron microscopy of the infection and subsequent development of soybean nodule cells, *J. Bacteriol.,* 92, 204, 1966.
16. **Newcomb, W., Sippel, D., and Peterson, R. L.,** The early morphogenesis of *Glycine max* and *Pisum sativum* root nodules, *Can. J. Bot.,* 57, 2603, 1979.
17. **Tu, J. C.,** Rhizobial root nodules of soybean revealed by scanning and transmission electron microscopy, *Phytopathology,* 65, 447, 1975.
18. **Vance, C. P., Johnson, L. E. B., Stade, S., and Groat, R. G.,** Birdsfoot trefoil (*Lotus corniculatus*) root nodules: morphogenesis and the effect of forage harvest on structure and function, *Can. J. Bot.,* 60, 505, 1982.
19. **Paau, A. S. and Cowles, J. R.,** Effect of induced nodule senescence on parameters related to dinitrogen fixation, bacteroid size and nucleic acid content, *J. Gen. Microbiol.,* 111, 101, 1979.
20. **Paau, A. S. and Cowles, J. R.,** Bacteroid distributions in alfalfa nodules upon dark-induced senescence and subsequent partial rejuvenation, *Physiol. Plant,* 52, 43, 1981.
21. **Roponen, I.,** The effect of darkness on the leghemoglobin content and amino acid levels in root nodules of pea plants, *Physiol. Plant.,* 23, 452, 1970.
22. **Arturri, P., Virtanen, I., and Laine, T.,** Red, brown and green pigments in leguminous root nodules, *Nature (London),* 157, 25, 1946.
23. **Lawrie, A. C. and Wheeler, C. T.,** The supply of photosynthetic assimilates to nodules of *Pisum sativum* L. in relation to the fixation of nitrogen, *New Phytol.,* 72, 1341, 1973.
24. **Butler, G. W., Greenwood, R. M., and Soper, K.,** Effects of shading and defoliation on the turnover of root and nodule tissue of *Trifolium repens, Trifolium pratense,* and *Lotus uliginosus, N. Z. J. Agric. Res.,* 2, 415, 1959.
25. **Whitman, P. C. and Lulham, L.,** Seasonal changes in growth and nodulation of perennial tropical pasture legumes in the field. I. The influence of planting date and grazing and cutting on *Desmodium uncinatum* and *Phaseolus atropurpureus, Aust. J. Agric. Res.,* 21, 195, 1970.
26. **Whiteman, P. C.,** Seasonal changes in growth and nodulation of perennial tropical pasture legumes in the field. II. Effects of controlled defoliation levels on nodulation of *Desmodium intortum* and *Phaseolus atropurpureus, Aust. J. Agric., Res.,* 21, 207, 1970.
27. **Dart, P. J. and Mercer, F. Y.,** The influence of ammonium nitrate on the fine structure of *Medicago tribuloides* Desr. and *Trifolium subterraneum* L., *Arch. Mikrobiol.,* 51, 233, 1965.
28. **Gibson, A. H. and Pagan, J. D.,** Nitrate effects on the nodulation of legumes inoculated with nitrate-reductase-deficient mutants of *Rhizobium, Planta,* 134, 17, 1977.
29. **Chen, P. C. and Phillips, D. A.,** Induction of root nodule senescence by combined nitrogen in *Pisum sativum* L., *Plant Physiol.,* 59, 440, 1977.
30. **Rigaud, J. and Puppo, A.,** Effect of nitrite upon leghemoglobin and interaction with nitrogen fixation, *Biochem. Biophys. Acta,* 497, 702, 1977.
31. **Manhart, J. R. and Wong, P. P.,** Nitrate effect on nitrogen fixation (acetylene reduction): activities of legume root nodules induced by rhizobia with varied nitrate reductase activities, *Plant Physiol.,* 65, 502, 1980.
32. **Groat, R. G. and Vance, C. P.,** Root nodule enzymes of ammonia assimilation in alfalfa (*Medicago sativa* L.): developmental patterns and response to applied nitrogen, *Plant Physiol.,* 67, 1198, 1981.
33. **Streeter, J. G.,** Synthesis and accumulation of nitrate in soybean nodules supplied with nitrate, *Plant Physiol.,* 69, 1429, 1982.

34. **Bassett, B., Goodman, R. N., and Novacky, A.,** Ultrastructure of soybean nodules. II. Deterioration of the symbioses in ineffective nodules, *Can. J. Microbiol.*, 23, 873, 1977.
35. **Werner, D., Morschel, E., Stripf, R., and Winchenbach, B.,** Development of nodules of *Glycine max* infected with an ineffective strain of *Rhizobium japonicum*, *Planta*, 147, 320, 1980.
36. **Groat, R. G. and Vance, C. P.,** Root and nodule enzymes of ammonia assimilation in two plant conditioned symbiotically ineffective genotypes of alfalfa (*Medicago sativa* L.), *Plant Physiol.*, 69, 614, 1982.
37. **Vance, C. P. and Johnson, L. E. B.,** Plant determined ineffective nodules in alfalfa (*Medicago sativa* L.): structural and biochemical comparisons, *Can. J. Bot.*, 61, 93, 1983.
38. **Klucas, R. V.,** Studies on soybean nodule senescence, *Plant Physiol.*, 54, 612, 1974.
39. **Kijne, J. W.,** The fine structure of pea root nodules. II. Senescence and disintegration of the bacteroid tissue, *Physiol. Plant Pathol.*, 7, 17, 1975.
40. **Swaraj, K. and Garg, O. P.,** The effect of aging on the leghemoglobin of cowpea nodules, *Physiol. Plant.*, 39, 185, 1977.
41. **Paau, A. S., Cowles, J. R., and Raveed, D.,** Development of bacteroids in alfalfa (*Medicago sativa*) nodules, *Plant Physiol.*, 62, 526, 1978.
42. **Vance, C. P., Heichel, G. H., and Barnes, D. K.,** Nitrogen fixation and nodule structure in perennial legumes, in *Proc. 8th N. Am. Rhizobium Conf.*, Clark, K. W. and Stephens, J. H. G., Eds., University of Manitoba Press, Winnipeg, 1981, 335.
43. **Sutton, W. D.,** Nodule development and senescence, in *Nitrogen Fixation, Vol. 3*, Broughton, W. J., Ed., Clarendon Press, Oxford, 1982, 144.
44. **Barker, K. R. and Hussey, R. S.,** Histopathology of nodular tissue of legumes infected with certain nematodes, *Phytopathology*, 66, 851, 1976.
45. **Orellana, R. G. and Worley, J. F.,** Cell dysfunction in root nodules of soybeans grown in the presence of *Rhizoctonia solani*, *Physiol. Plant Pathol.*, 9, 183, 1976.
46. **Tu, J. C.,** Effects of soybean mosaic virus infection on ultrastructure of bacteroidal cells in soybean root nodules, *Phytopathology*, 67, 199, 1977.
47. **Orellana, R. G. and Fan, F.,** Nodule infection by bean yellow mosaic virus in *Phaseolus vulgaris*, *Appl. Environ. Microbiol.*, 36, 814, 1978.
48. **Orellana, R. G., Fan, F., and Sloger, C.,** Tobacco ringspot virus and *Rhizobium* interactions in soybean: impairment of leghemoglobin accumulation and nitrogen fixation, *Phytopathology*, 68, 577, 1978.
49. **Sprent, J. I.,** Nitrogen fixation by legumes subjected to water and light stresses, in *Symbiotic Nitrogen Fixation in Plants*, Nutman, P. S., Ed., Cambridge University Press, London, 405, 1976.
50. **Minchin, F. R. and Summerfield, R. J.,** Symbiotic nitrogen fixation and vegetative growth of cowpea (*Vigna unguiculata* L. Walp.) in waterlogged conditions, *Plant Soil*, 45, 113, 1976.
51. **Pate, J. S.,** Functional biology of dinitrogen fixation by legumes, in *A Treatise on Dinitrogen Fixation, Section III: Biology*, Hardy, R. W. F. and Silver, W. S., Eds., John Wiley & Sons, New York, 473, 1977.
52. **Bisseling, T., van Staveren, W., and van Kammen, A.,** The effect of waterlogging on the synthesis of the nitrogenase components in bacteroids of *Rhizobium leguminosarum* in root nodules of *Pisum sativum*, *Biochem. Biophys. Res. Commun.*, 93, 687, 1980.
53. **Sacher, J. A.,** Senescence and post harvest physiology, *Annu. Rev. Plant Physiol.*, 24, 197, 1973.
54. **Noel, D. K., Carneol, M., and Brill, W. J.,** Nodule protein synthesis and nitrogenase activity of soybeans exposed to fixed nitrogen, *Plant Physiol.*, 70, 1236, 1982.

Chapter 8

# CHLOROPLAST SENESCENCE AND PROTEOLYTIC ENZYMES

## Michael J. Dalling and Angela M. Nettleton

### TABLE OF CONTENTS

| | | |
|---|---|---|
| I. | Introduction | 126 |
| II. | Characteristics of Chloroplast Senescence | 126 |
| | A. Change in Chloroplast Number | 126 |
| | B. Changes in Chloroplast Ultrastructure | 127 |
| |     1. Natural Senescence | 127 |
| |     2. Induced Senescence | 130 |
| |         a. Leaf Detachment | 130 |
| |         b. Growth Regulators | 131 |
| | C. Changes in Chloroplast Composition | 131 |
| |     1. Pigments and Pigment-Protein Complexes | 131 |
| |     2. Lipids | 134 |
| | D. Changes in Chloroplast Function | 134 |
| |     1. $CO_2$-Assimilation | 134 |
| |     2. Electron Transport | 135 |
| |     3. Photophosphorylation | 135 |
| |     4. Capacity for Protein Synthesis | 136 |
| III. | Degradative Events — Differentiation and Senescence | 137 |
| | A. Vacuolar Influences | 137 |
| |     1. Pinocytosis | 137 |
| |     2. Change in Transfer Properties of Tonoplast | 138 |
| | B. The Autonomous Nature of Chloroplasts | 138 |
| |     1. Chlorophyll Degradation | 139 |
| |     2. Lipid Degradation | 139 |
| |     3. Nucleic Acid Degradation | 139 |
| |     4. Protein Degradation | 140 |
| |         a. Light Harvesting Chlorophyll *a/b* Protein | 140 |
| |         b. $Q_B$-Protein (Peak D, Photogene 32) | 141 |
| |         c. NADPH 3-Protochlorophyllide Oxidoreductase | 141 |
| |         d. Ribulose 1,5-Biphosphate Carboxylase | 142 |
| IV. | Role of Proteolytic Enzymes | 142 |
| | A. Influence of Nuclear and Chloroplast Genomes | 142 |
| | B. Compartmentation of Proteolysis | 143 |
| |     1. Degradation of Thylakoid Proteins | 143 |
| |         a. Speculation No. 1 | 143 |
| |         b. Speculation No. 2 | 144 |
| |     2. Degradation of Stromal Proteins | 145 |
| | C. Initiation of Proteolysis | 145 |
| |     1. Transcription of Nuclear DNA | 145 |
| |     2. Targeting of Protein Substrates | 145 |

3. Activation by Unsaturated Fatty Acids..........................145

References.......................................................................146

## I. INTRODUCTION

The process of photosynthesis is arguably the most important process of biology. Not only does this process replenish the oxygen content of the atmosphere, it also converts carbon dioxide to sugars. The driving force for these biochemical reactions is energy derived from a complex system of light harvesting pigments in the plant. All these reactions — $O_2$-evolution, $CO_2$-assimilation, and light harvesting — take place in the chloroplast.

All agricultural produce is derived from photosynthesis, harvested either as a somewhat direct product such as forage and grain or, as an indirect product such as milk and meat. Productivity can, therefore, be considered in terms of maximizing photosynthesis per unit land area. This ambition has many levels of expression, but one that is the most poorly understood is delaying the deterioration or senescence of the chloroplast. Senescence may occur as a natural consequence of plant ontogeny or may be induced by factors such as low nutrition, water stress, or disease. In any case, photosynthesis declines and, as a consequence, so does productivity.

In spite of the importance of chloroplasts, and in contrast to our extensive knowledge of the processes associated with photosynthesis, we know little of the mechanisms responsible for senescence of this organelle.

## II. CHARACTERISTICS OF CHLOROPLAST SENESCENCE

### A. Change in Chloroplast Number

The earliest recorded studies of chloroplast senescence are those of Mohl[1] in 1845 who examined the nature of winter browning of conifers. In 1885 Schimper[2] reported on the structure of chloroplasts in both algae and vascular plants and concluded that the change in color of leaves during the autumn was due to degradation of chlorophyll and was not associated with destruction of the chloroplast. Schimper proposed that changes in plastid structure are not unidirectional or nonreversible, but rather, they may often follow a cyclic pattern. This concept has received considerable support,[3] and the best documented example is the formation of chromoplasts.[3,4] The alternate view proposes that chloroplast development is unidirectional with chloroplast senescence an irreversible phenomenon that ultimately results in destruction or loss of the plastid.[5]

The question as to whether chloroplast number declines as a consequence of senescence has been the topic of many subsequent studies. Other early studies of winter browning in evergreen species[6-8] concluded that the chloroplasts were lost in autumn or early winter. However, more recent studies with improved methods of tissue fixation strongly suggest that in evergreen leaves the chloroplasts may change in appearance in response to low winter temperatures, but they do not disintegrate.[8-11]

The terminal phase of senescence, coincident with leaf fall in deciduous plants such as *Betula*[12] and *Larix decidua* × *kaempferi*[13] is, however, characterized by complete loss of chloroplast integrity. Similarly, it would appear that there is complete disintegration of chloroplasts in the mesophyll cells of senescent leaves of *Taxus baccata* L., an evergreen gymnosperm in which the leaves typically have a life span of 4 to 5 years.[14]

Direct counting of the chloroplast population of mesophyll cells during leaf senescence has produced conflicting results, some of which can be attributed to differences in techniques.[15] Wittenbach et al.[16] and Lamppa et al.[17] observed a reduction in chloroplast number per mesophyll cell during senescence of wheat and pea leaves, respectively. In contrast, Martinoia et al.[18] and Wardley et al.[15] observed no change in chloroplast number per mesophyll cell during senescence of barley and wheat leaves, respectively. During the course of senescence of the primary leaf of wheat,[15] chlorophyll and the chloroplast enzyme ribulose-1,5-bisphosphate carboxylase (RuBPCase), declined by more than 80%. The number of chloroplasts per mesophyll cell did not change appreciably, although late in senescence there was a small decline, particularly in the mesophyll cells of the leaf tip. A reduction in chloroplast number is probably attributable to events associated with the terminal stages of senescence, occurring after the major portion of the mobilizable elements are lost from the plastid.

## B. Changes in Chloroplast Ultrastructure
### 1. Natural Senescence

The first detectable symptoms of leaf senescence are usually expressed in the chloroplast. There have been many studies of chloroplast ultrastructure during the course of leaf senescence,[12,19-28] and collectively these indicate a distinctive and predictable sequence of changes (Figures 1 to 6). The appearance of the thylakoids changes first as they become dilated and disorganized (Figures 1 and 2). Initially the stroma lamella are lost in preference to the granal lamellae (Figure 1 cf., Figures 2 and 3). Ultimately, the entire lamellar system of the thylakoids is reduced to membrane vesicles or membrane sheets (Figure 4). Coincident with some of these changes is a degradation of the stromal proteins and an increase in amount of plastoglobuli (Figure 1 cf., Figure 3). The chloroplast envelope remains intact until major degradation of the stroma and thylakoid lamellae has occurred.[26,28] Peoples et al.[26] observed the outer component of the envelope double membrane to rupture before the inner component. At very advanced stages of cell senescence the senesced chloroplasts persisted even though the other organelles had degenerated.[12,24]

Mittelheuser and Van Steveninck[29] observed that one of the earliest events to occur during chloroplast senescence was the loss of recognizable ribosomes, which occurred before degradation of the cytoplasmic ribosomes. This observation is in accord with biochemical analyses of chloroplast senescence (Section II.D.4).

Dodge[12] has documented some of these changes in a quantitative manner to describe chloroplast ultrastructure during autumnal senescence of *Betula* leaves (Figure 7). The shape of the chloroplast changes from elliptical to spherical (Figure 7A and B), an observation recorded in most other studies. The chloroplast volume declines markedly (Figure 7C), due no doubt to the degradation of the stroma and thylakoids and the export of some breakdown products from the chloroplast. Martinoia et al.[18] found that the buoyant density of chloroplasts isolated from the primary leaf of barley declined from 1.208 g/cm$^3$ to 1.186 g/cm$^3$ as senescence progressed. This suggests that within the chloroplast, lipids, or the breakdown products of lipid degradation, are retained more than the breakdown products of protein degradation.

Lichtenthaler[30] has proposed that plants can be divided into two groups on the basis of whether the number of plastoglobuli increases or remains constant during senescence. *Betula* clearly belongs to the latter group.[12] Annual plants such as wheat belong to the former group.[26,28]

Plastoglobuli, or osmiophilic granules, are lipid inclusions that occur in the stroma of all chloroplasts and have no direct association with the thylakoids. Lipophilic plastidquinones (plastoquinone 45, plastoquinol, α-tocopherol, α-tocoquinone) are the main lipids of isolated plastoglobuli.[30] During thylakoid degeneration, the plastoglobuli also accumulate carotenoids.[30-32] The increase in size and frequency of plastoglobuli during chloroplast senescence,

**128** *Plant Proteolytic Enzymes*

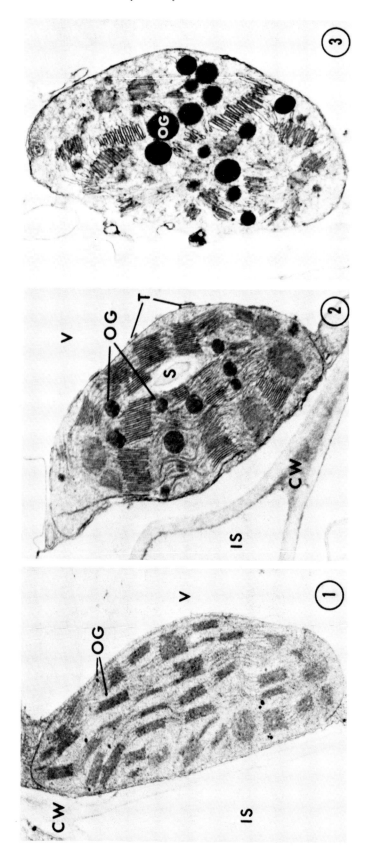

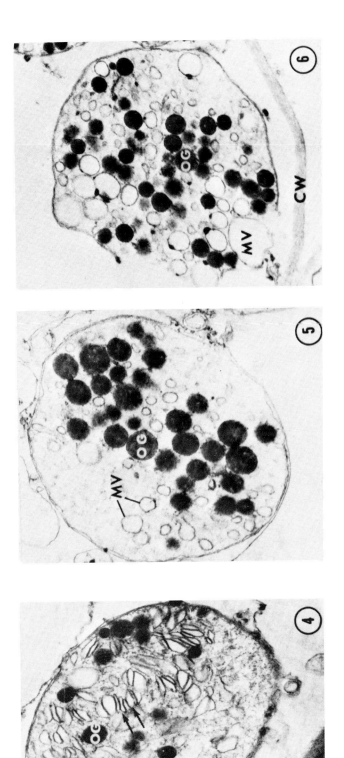

FIGURES 1 to 6. Chloroplasts of the primary leaf of wheat at different stages of senescence.[28] (1): 10-day-old leaf. Chloroplasts contain small asmiophillic globuli (OG) and a well organized thylakoid system. CW, cell walls; IS, intercellular space; V, vacuole. (2) and (3): 21-day-old leaf, green basal region. Chloroplasts contain large osmiophillic globuli; grana and intergranal lamellae are swollen (Figure 2) or lamellar system disoriented and grana narrower. T, tonoplast fragments. (4): 21-day-old leaf, yellow-green mid-region. Number of osmiophillic granules has increased; thylakoid lamellae have re-oriented to form appressed membrane vesicles (arrow). (5) and (6): 21-day-old leaf, yellow apex. Numerous membrane vesicles (MV) and large osmiophillic globuli.

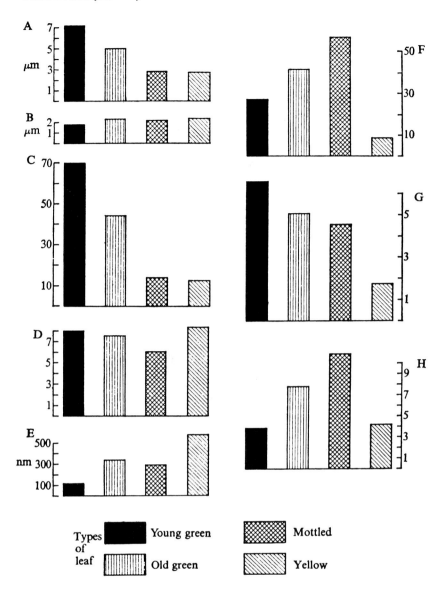

FIGURE 7. Histograms summarizing the main changes in chloroplast size and composition during autumnal senescence of *Betula* leaves. A, length; B, width; C, approximate volume ($\mu m^3$); D, number of plastoglobuli per section; E, average size of plastoglobuli; F, number of thylakoids in a vertical transect through the center of a chloroplast; G, number of grana per vertical section of chloroplast; H, number of thylakoids per granum. (From Dodge, J. D., *Ann. Bot.*, 34, 817, 1970. With permission.)

and their closeness to degenerating membranes led Lichtenthaler and Weinert[33] to the suggestion that plastoglobuli contain lipids and carotenoids released from degraded thylakoids. The accumulation of lipophilic material in the plastoglobuli undoubtedly contributed to the decline in chloroplast buoyant density referred to previously.

### 2. Induced Senescence
#### a. Leaf Detachment

Detachment of leaves and incubation in the dark induces rapid senescence.[34-37] In many respects, the changes in chloroplast ultrastructure[22,25,28,29,37-41] follow those described for attached leaves senescing naturally (Section II.B.1). Hurkman[28] has made a detailed comparison of chloroplast ultrastructure during the course of senescence by attached and detached wheat leaves. He concludes that the initial stages of chloroplast degeneration are very similar — thylakoid lamellae are distorted, plastoglobule formation is increased, and the stroma is degraded. However, due possibly to the greatly accelerated rate of senescence, other changes are not observed For example, the disruption of the thylakoid lamellae is not as severe, and fewer plastoglobuli are formed. In addition, the chloroplast envelope is ruptured well before degradation of the thylakoid lamellae. Hurkman[28] and Colquhoun et al.[40] correctly conclude that great care should be taken in extending observations made on detached, senescing leaves to the development of hypotheses concerning the mechanism and control of senescence of attached leaves.

### b. Growth Regulators

The consequences of applying kinetin to senescing leaves has been investigated in many laboratories.[34,42-54] The general consensus is that kinetin delays, but does not prevent, protein and chlorophyll loss. There have been several studies conducted to investigate the effect that kinetin application has on chloroplast ultrastructure.[22,29,37,49,52,54] Kinetin application retards the loss of chloroplast ribosomes[29,37,39,55,56] and delays chlorophyll loss, possibly by stabilizing the grana.[37,39,49,52,55,57]

Exposure of plant tissues to ethylene accelerates senescence,[58] a phenomenon of great practical and economic importance in the horticulture and floriculture industries.[59] Ethylene induces the formation of macrograna in chloroplasts of attached leaves of tomato[54] and Morning Glory.[60] This symptom parallels the effect of inhibition of protein synthesis on 70S ribosomes,[60-62] and the effect of high temperature,[63] some mineral deficiencies,[64] and genetic inhibition of chloroplast development.[65] Phyto-ferritin paracrystalline structures accumulate in the chloroplasts of Morning Glory leaves exposed to ethylene,[66] a response apparently common to several other treatments that induce senescence.

## C. Changes in Chloroplast Composition
### 1. Pigments and Pigment-Protein Complexes

Free chlorophyll does not exist in the chloroplast: it is found complexed with specific proteins that act to hold the pigment within the thylakoid.[67] There are three major chlorophyll-protein complexes: the light harvesting chlorophyll $a/b$ protein (LHCP): and the two photosystems, PS I containing $P_{700}$ and PS II containing $P_{680}$. The LHCP functions as an antenna for PS II, and since it contains most of the chlorophyll $b$, has a chlorophyll $a/b$ ratio near 1.[68] LHCP can account for 50% of the chlorophyll, PSI about 30%, and the balance is associated with PS II.[69] The chlorophyll-protein complexes are not evenly distributed in the thylakoids. The PS II and LHCP are located in the grana, and PS I is associated with the stroma lamellae.[70,71]

The most conspicuous change associated with leaf senescence is the loss of pigments, especially chlorophyll[72] (Figure 9A). Direct measurement of the chlorophyll content of individual chloroplasts shows clearly that chlorophyll is lost from individual chloroplasts (Figure 8), rather than through a reduction in chloroplast number per mesophyll cell (Section II.A).

Biswal et al.[73] have reported on the comparative rates of breakdown of carotenoids and chlorophylls of mustard seedlings during dark-mediated senescence. Carotenoids persist longer than chlorophyll in senescing leaves,[73,74] and this difference manifests itself in leaf yellowing. By contrast, Panigrahi and Biswal[75] observed more rapid loss of carotenoids than chlorophyll during incubation of isolated chloroplasts. The retention of carotenoids in mustard seedlings is proposed by Biswal et al.[73] to be attributed to the greater lipid solubility of carotenoids. They postulate that upon release from the membrane during thylakoid disas-

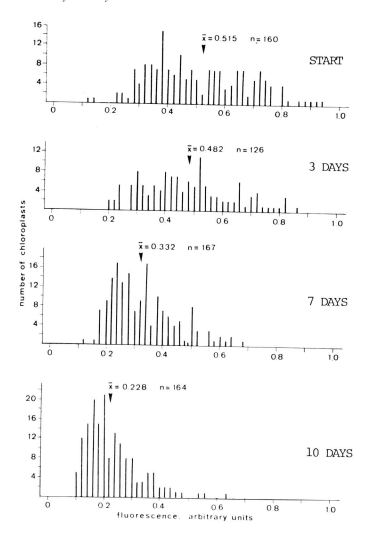

FIGURE 8. Histograms showing the relative chlorophyll content of individual chloroplasts isolated from the primary leaf of barley induced to senesce for 0, 3, 7, and 10 days in the dark. (Adapted from Martinoia, E., Heck, U., Dalling, M. J., and Matile, Ph., *Biochem. Physiol. Pflanzen.*, 178, 147, 1983. With permission.)

sembly, the carotenoids are immediately enclosed and protected by lipid globules, which are formed during the breakdown of chloroplasts (Section II.B.1).

During senescence of the chloroplast, not all of the thylakoid polypeptides are lost at the same rate. Bricker and Newman[76] recognized two groups of proteins, those that were degraded rapidly and those more resistant to hydrolysis. The earliest indication that the chlorophyll:protein complexes were lost at different rates came from Sestak,[77] who found a higher ratio of PS I/PS II particles in young leaves of spinach compared with older leaves. This has been confirmed in several subsequent studies. Sestak et al.[78] observed that during senescence of the primary leaf of *Phaseolus vulgaris* L. the ratio of PS I/PS II declines. Bricker and Newman[79] isolated the protein:chlorophyll complexes by SDS-PAGE. They observed that during senescence of *Glycine max* L. cotyledons the amount of $P_{700}$ Chl *a*-protein complex (PS I) declined more rapidly than the protein:chlorophyll complexes associated with PS II, including LHCP. This difference was reflected by a more rapid loss of

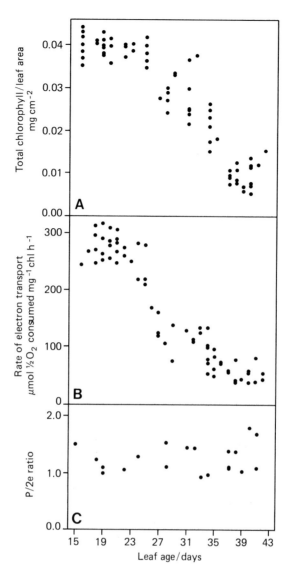

FIGURE 9. Some characteristics of senescence of the primary leaves of *Phaseolus vulgaris* L. (Adapted from Jenkins, G. I. and Woolhouse, H. W., *J. Exp. Bot.*, 32, 467, 1981. With permission.)

electron transport activity associated with PS I. Jenkins et al.[80] resolved the protein:chlorophyll complexes from senescing leaves of *P. vulgaris* L. and also found a faster rate of decline for the $P_{700}$ Chl *a*-protein complex. They confirmed this observation through age-related changes in the 77 K fluorescence emission spectrum at 685, 694, and 734 nm, corresponding to LHCP, PS II, and PS I, respectively. While acknowledging the difficulty of making quantitative conclusions from these data, they conclude that the changes in fluorescence emission at 734 nm are consistent with selective loss of PSI.

In a nonyellowing mutant of *Festuca pratensis*, Thomas[81] found that some thylakoid proteins were degraded more readily than others during induced senescence. Susceptibility to degradation was not an intrinsic feature of the thylakoid proteins, because once they had been extracted from the thylakoids, all the proteins were degraded by a crude proteolytic enzyme preparation.[82] Susceptibility to degradation appears to be modulated by accessibility to the particular peptide hydrolase (Section IV.B.1). In the mutant, some thylakoid proteins, especially those complexed with chlorophyll, are less accessible than those of the wild-type genotype.

Degradative changes associated with senescence of the thylakoid polypeptides appear to focus initially on a small group of critical proteins, e.g., $P_{700}$ Chl $a$-protein complex and the Cyt f/$b_6$-protein complex (Section II.D.2). Whether this is due entirely to the accessibility of these proteins is still not clear. However, it is interesting to note that the ultrastructural studies of chloroplast senescence support this hypothesis to some extent. The stroma lamellae, which are the primary site of PS I, are disassembled well before the granal lamellae (Section II.B.1). Thus stacking itself may provide a physical barrier to access by an attacking protease, and the differential location of the components of the whole photochemical apparatus may be an important determinant of their susceptibility to degradation.

*2. Lipids*

The most abundant lipids of the chloroplast are monogalactosyldiacylglycerol (MGDG) and digalactosyldiacylglycerol (DGDG).[83] The major portion of the chloroplast galactolipid is associated with the thylakoid lamellae. MGDG is the most abundant lipid of the thylakoid membrane and DGDG the predominant envelope galactolipid. The other important structural lipids are sulfolipid, sterols, and the phospholipids: phosphatidylcholine and phosphatidylethanolamine. During senescence of the chloroplast, changes in the level of these compounds are readily detected. Galactolipids, sulfolipids, and phospholipids decline in parallel as senescence proceeds.[84-87] Total sterol content of the thylakoids does not change, and there is no significant qualitative change in the fatty acid composition of the extractable membranes.[88]

One important consequence of the changes in the chemical composition of the chloroplast (thylakoid) membranes, is a shift in the physical state of the membrane lipids. Wide-angle X-ray diffraction studies indicate that as senescence proceeds, membranes acquire increasing proportions of gel phase lipid. During senescence of the primary leaf of *P. vulgaris* L. the transition temperature for thylakoid membranes from young leaves is about $-30°C$, indicating that at physiological temperatures the lipid is entirely liquid-crystal.[88] The lipid phase transition temperature for chloroplasts from leaves that had lost over half their protein and chlorophyll was greater than 30°C. McKersie and Thompson[88] suggest that the increase in free sterols and concomitant decline in chlorophyll and thylakoid polypeptides (Section II.C.1) cause a redistribution of polar lipids that results in the formation of the gel phase. Whether the change in physical state of the thylakoid membrane is the cause of membrane senescence, or simply a consequence of other deteriorative events, is uncertain.

### D. Changes in Chloroplast Function

*1. $CO_2$-Assimilation*

One of the most important features of leaf senescence is a decline in the rate of $CO_2$-assimilation.[27,89,90] This has often been attributed to an increase in mesophyll resistance,[89] or more specifically, a decline in RuBPCase content or activity.[89-91] However, the activity of other stromal enzymes, e.g., NADP-triose-phosphate dehydrogenase,[92,93] ribulose 5-phosphate kinase,[93] 3-phosphoglyceric acid kinase,[93] and fructose 1,6-bisphosphatase[94] also decline in parallel to RuBPCase. In addition, the rate of electron transport (Section II.D.2) and photophosphorylation (Section II.D.3) decline during senescence. One consequence of these changes would be a reduction in the rate of regeneration of ribulose 1,5-bisphosphate, and this could reduce $CO_2$-assimilation independently of the activity of RuBPCase.

Evans[95] observed that as wheat leaves aged the calculated Hill activity and RuBPCase activity decline in parallel. Therefore, it would appear that neither RuBPCase activity nor the rate of electron flow are solely responsible for the decline in $CO_2$-assimilation during senescence. Rather, it would seem that degradation of the photochemical and biochemical components of photosynthesis occurs in a coordinated manner.

### 2. Electron Transport

Sestak[96] has made an extensive review of the changes in photosynthetic electron transport during leaf ontogeny. Although these studies only documented changes in Hill activity, collectively they do show that leaf senescence is associated with a loss of photosynthetic electron transport capacity.

Several aspects of the biochemical basis for the decline in photosynthetic electron transport have been examined. One line of investigation has considered free fatty acids, which are reputed to have a deleterious effect on the structure and function of isolated chloroplasts. These observations have given rise to the hypothesis that senescence of chloroplasts within intact tissue may be modulated or enhanced by free fatty acids released as a consequence of lipid degradation.[98] One phenomenon that has been reported extensively is the inhibition by free fatty acids of the in vitro rate of electron transport by thylakoids.[97,99-104] The relevance of these in vitro experiments has been questioned by Percival et al.,[97] who observed no correlation between the loss of Hill activity and the release of linolenic acid from endogenous lipid in *Vicia faba* chloroplasts aged under different conditions. In fact, it was necessary for the endogenous level of linolenic acid to be about tenfold higher than the level *in situ* if there was to be any appreciable effect on Hill activity. It would seem, therefore, that an alternate explanation is necessary.

Jenkins and Woolhouse[105] measured the activities of PS I and PS II in chloroplasts isolated from the primary leaves of *P. vulgaris* L. They found that during senescence the rates of electron flow through PS I and PS II declined by 25 and 33%, respectively. However, the rate of coupled, noncyclic electron transport declined by 80% during the same period (Figure 9B). This disparity in rates of electron transport indicates that an impairment of electron flow between PS I and PS II was responsible for the decline in coupled, noncyclic electron transport.[106] Further studies revealed that the rate of electron flow from the plastoquinone pool to PS I, rather than transfer into the plastoquinone pool from PS II, declined markedly during senescence.[107]

Holloway et al.[108] have considered the changes in photosynthetic electron transport during senescence of *Hordeum vulgare* L. leaves in terms of three independent complexes — PS I, PS II, and the Cyt f/$b_6$-protein complex, linked by the mobile electron carriers plastocyanin and plastoquinone. Senescence was associated with a decline in the concentration of cytochromes f and $b_6$ (Figure 10), and as a consequence, the rate of oxidation of plastohydroquinone was limited. The critical role of the Cyt f/$b_6$-protein complex has been confirmed by Ben-David et al.[109] Using an immunological approach, these investigators observed a significant decline in the polypeptides associated with the Cyt f/$b_6$-protein complex during senescence of *P. vulgaris* L. and *Avena sativa* L. leaves. In contrast, there was little change during the same period in the amount of Coupling Factor, a complex protein that couples electron transport to ATP synthesis.

In conclusion, it would appear that the decline in photosynthetic noncyclic electron transport during senescence is due, at least in the first instance, to a selective degradation or removal from the thylakoid membrane of the Cyt f/$b_6$-protein complex.

### 3. Photophosphorylation

Despite an 80% decline in the rate of coupled, noncyclic electron transport (Figure 9B), this reaction remained coupled to photophosphorylation with an average P/2e ratio of about 1.3 during the life span of the leaf (Figure 9C). This result is consistent with the observation by Ben-David et al.[109] that Coupling Factor was not lost during senescence.

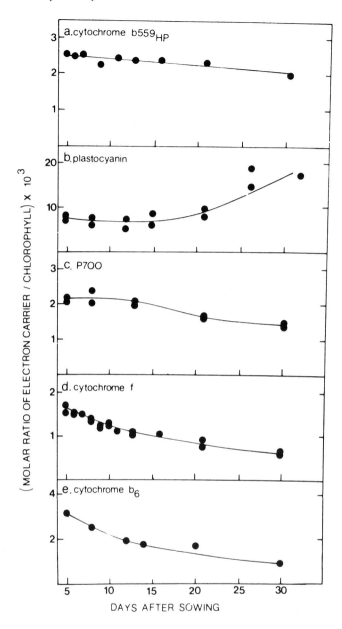

FIGURE 10. Concentrations of five electron carriers in thylakoids isolated from barley leaves of different ages. (From Holloway, P. J., Maclean, D. J., and Scott, K. J., *Plant Physiol.*, 72, 795, 1983. With permission.)

### 4. Capacity for Protein Synthesis

Because of its great abundance in the chloroplast, RuBPCase was quickly adopted as a useful model to study the control of protein synthesis. RuBPCase has a molecular weight of about 525 kdaltons and has two types of subunits: large subunit (LSU) of 55 kdaltons and small subunit (SSU) of 13 kdaltons.[110-112] RuBPCase consists of eight subunits of each type.[113] The LSU is synthesized on chloroplast ribosomes,[114] and the SSU is synthesized on cytoplasmic ribosomes[115] as a precursor protein of molecular weight about 20 kdaltons.[116] The precursor protein is processed to release the mature SSU of molecular weight 13 kdaltons when it enters the chloroplast (Chapter 4).

The synthesis of RuBPCase occurs at a rate similar to that of other proteins as leaves approach full expansion.[117] After full expansion, the rate of RuBPCase synthesis declines relative to the rate of synthesis of other cellular proteins.[118-122] This decline is due to a coordinated reduction in synthesis of the LSU and SSU.[117]

The decline in chloroplast protein synthesis during leaf senescence can be attributed to several factors: the amount of chloroplast r-RNA declines relative to cytoplasmic r-RNA;[119,120,123] the number of polysomes in the chloroplast also declines.[119,123] The rate of decline of RuBPCase synthesis, however, exceeded the decline in the number of chloroplast polysomes.[124] In the cytoplasm, the number of polyribosomes is maintained, at least until late in senescence.[119,123,124]

Spiers and Brady[125] measured the amounts of translatable mRNA for the LSU and SSU of RuBPCase during senescence of the second leaf of wheat seedlings. They observed that both mRNAs declined dramatically relative to most of the other mRNAs.

The decline in chloroplast RNA is due at least in part to reduced synthesis. For example, Callow[120] observed that the incorporation of petiole-fed $^{32}$P into chloroplast rRNA takes place only in young expanding leaves. In contrast, label continued to be incorporated into cytoplasmic rRNA after the leaves reached full expansion. Ness and Woolhouse[126] measured the in vitro activity of chloroplast DNA-dependent RNA polymerase during growth and senescence of *P. vulgaris* L. leaves. The rate of RNA synthesis per leaf began to fall before the leaf was 70% expanded, and by full expansion chloroplast RNA polymerase activity had declined by 70%. The rate of chloroplast RNA synthesis per unit chlorophyll was declining before the leaves reached 25% expansion.

The role of chloroplast ribonuclease enzymes during senescence has not been determined. These enzymes are readily detected in chloroplasts and are presumed to be involved with nucleic acid turnover (Section III.B.3).

## III. DEGRADATIVE EVENTS — DIFFERENTIATION AND SENESCENCE

### A. Vacuolar Influences

It has been hypothesized that the vacuole in higher plants functions in a manner similar to lysosomes in animal cells.[127] The detection of hydrolytic enzymes in isolated vacuoles by several workers gives support to this hypothesis. Boller and Kende,[128] Waters et al.,[129] and Noble and Dalling[130] have reported the vacuolar localization of acid phosphatase, α-mannosidase, *N*-acetyl-β-D-glucosaminidase, and phosphodiesterase. Boller and Kende[128] have also detected β-fructosidase activity. Acid phosphatase and nuclease activity in vacuoles has been reported by Butcher et al.[131] and Buser and Matile.[132] RNase activity has been identified by Baumgartner and Matile.[133] Most reports on the localization of vacuolar enzymes have included the recognition of proteolytic activity in vacuoles; these include both exo- and endopeptidases.[128-130,134-137]

The idea that control of degradation can be achieved through separation of the various hydrolases and their respective substrates by intracellular compartmentation is simple enough in principle. But how selectivity of degradation is achieved has clearly tested the ingenuity of the proponents of the scheme. Two hypotheses have been advanced, and these are outlined below.

*1. Pinocytosis*

According to this hypothesis, the vacuole and chloroplast interact in such a way that the vacuole engulfs whole or fragmented chloroplasts and their constituent proteins, lipids, pigments, and nucleic acids, and other compounds are then degraded by the vacuolar hydrolases. It would follow from this hypothesis that senescence is accompanied by a reduction in chloroplast number per cell. The data of Wittenbach et al.[16] and, to a lesser extent Lamppa et al.[17] are consistent with this hypothesis. However, the ultrastructural evidence presented by Wittenbach et al.[16] is at variance with the many previous studies on this topic (Section II.B.1). By contrast, other studies indicate that chloroplasts exhibit a high degree of intactness, despite substantial dissolution of the stroma and thylakoids. Studies that have examined the changes in chloroplast number during senescence give only tenuous support to the pinocytotic hypothesis (Section II.A.). Recent studies, using different techniques,[15,18] both show unequivocally that there is no significant change in chloroplast number during the course of mesophyll cell senescence. The third line of evidence that opposes pinocytosis as the mechanism by which chloroplasts senesce, comes from the studies of Thomas with the nonyellowing mutant of *Festuca pratensis*.[138] A considerable distinction exists among proteins of the thylakoid membrane of this mutant with respect to their degradation during senescence. The extrinsic proteins of the thylakoids and the stromal proteins are apparently degraded in a manner comparable to those of the wild-type genotype.[81,139] The intrinsic proteins, in particular the LHCP and other chlorophyll:proteins, are retained, resulting in the nonyellowing appearance of the mutant. Such discrimination would seem to oppose totally the pinocytotic hypothesis. Ultrastructural examination of the senescing tissue failed to reveal the existence of any chloroplasts, or chloroplast fragments, within the vacuole.[139] This finding was subsequently confirmed by direct counting of the chloroplasts during the course of senescence.[140]

The inadequacy of the pinocytotic hypothesis is further illustrated by the unusual case of the dessication tolerant plant *Borya nitida* Labill. During humidity-sensitive degreening of this plant, Hetherington et al.[141] showed that dehydration paralleled loss of chlorophyll, chloroplast proteins, and degradation of thylakoid membranes. However, the chloroplasts remained intact and could be restored to functional entities after rehydration. These observations suggest that the degradation of chloroplastic proteins is not due to general deterioration of that organelle, but rather to selective protein loss.

*2. Change in Transfer Properties of Tonoplast*

In many respects, this hypothesis can be seen as an extension of the pinocytotic or autophagic vacuole hypothesis and in principle has much in common with similar models described for animal cells. Specificity of degradation according to this hypothesis would be achieved by selective movement of particular or targeted compounds into the vacuole. This hypothesis has been examined in *Lemna minor* by Cooke et al.[142] who detected stress-induced changes in the permeability to amino acids of the tonoplast. These authors speculate that the increase in protein degradation induced by $NO_3$-starvation is a consequence of increased protein transfer across the tonoplast. Such a scheme must by its very nature be complex. A cytoplasm-vacuole model is quite complicated, but a chloroplast-vacuole model, requiring that both the double membrane of the chloroplast envelope, and the single membrane of the tonoplast be traversed, would seem to be unnecessarily complicated.

## B. The Autonomous Nature of Chloroplasts

An extensive literature establishes that chloroplasts contain the four components necessary for autonomy, that is, DNA, DNA polymerase, RNA polymerase, and protein-synthesizing system. However, it is also clear that the DNA does not code for all the chloroplast proteins, nor does the protein-synthesizing system make all of the chloroplast proteins.[143,144] The function and development of chloroplasts must, therefore, result from an interplay between chloroplast and nuclear genomes (see Chapter 4).

The impact that the nuclear-chloroplast genome interaction has on senescence of the chloroplast is not clear. In particular, it is not known whether the necessary hydrolytic enzymes are encoded by nuclear or chloroplast DNA. In spite of this important gap in our knowledge, there is an increasing literature which clearly established that the chloroplast contains a wide spectrum of hydrolytic enzymes. While these enzymes clearly possess the necessary hydrolytic activity in vitro, their role, if any, during chloroplast senescence has yet to be established. Their presence in the chloroplast, together with the data reviewed previously, strongly supports the idea that senescence of chloroplasts is a deteriorative process that occurs within the intact organelle. In this sense, the chloroplast might reasonably be considered to be autonomous.

*1. Chlorophyll Degradation*

Chlorophyll degradation has been described extensively as the first visible symptom of leaf senescence (Section II.C.1), and yet little is known of the processes responsible for this stage of development.

Two enzymes have been shown in vitro to be capable of chlorophyll degradation — chlorophyllase, which removes the phytol side chain with the production of free phytol and chlorophyllide,[145,146] and a peroxidase.[147,148] It is difficult to assign a role for the peroxidase *in situ*. Its location within the vacuole is consistent with the role that this organelle plays as a depository of deteriorative enzymes, but, like the other enzymes of the vacuole, there is great difficulty in equating in vitro activity and *in situ* function. The case for chlorophyllase having an active role in chlorophyll degradation seems stronger,[145] but unproven.

Martinoia et al.[98] have demonstrated the existence in mature chloroplasts of peroxidative and oxidative enzymes that catalyze the bleaching of chlorophyll. The activities of both enzymes are latent. They suggest that the requirement of detergent solubilization or sonication of the membranes for activation of the peroxidase, and addition of unsaturated fatty acids for activation of the oxidase, reflects the natural tendency for these enzymes to become active upon disorganization of the thylakoids and hydrolysis of the membrane lipids during senescence.

*2. Lipid Degradation*

In 1966, Barton[23] reported that the accumulation of osmiophilic globular bodies in senescing chloroplasts indicated a build-up of lipid material originating from thylakoid breakdown, this was supported by work of Lichtenthaler and Weinert[33] (Section II.B.1). The most abundant lipids of the chloroplast are the galactolipids MGDG and DGDG (Section II.C.2). Hydrolysis of these compounds to their constitutive glycerol, galactose, and free fatty acid components requires the concerted effort of three enzymes: α-galactosidase, β-galactosidase, and galactolipase. The role that each enzyme plays in the degradation of MGDG and DGDG is illustrated in Figure 11. Sastry and Kates[149] were the first to report that enzymes associated with chloroplasts from *P. multiflorus* were capable of releasing linolenic acid and the respective mono- or digalactosylglycerol from monogalactosyldilinolenin and digalactosyldilinolenin. Addition of a crude extract to this system caused further degradation of the monogalactosylglycerol and digalactosylglycerol to glycerol and galactose. The activity of a galactolipase can also be assumed from several reports in the literature that demonstrate a significant time dependent increase in the free fatty acid content of isolated chloroplasts.[97,101,150-152]

A galactolipase has been partially purified from chloroplasts of *P. vulgaris*,[153] and recently, a distinct stromal form of the enzyme β-galactosidase was described for wheat leaves.[154] No α-galactosidase has yet been reported in isolated chloroplasts.

*3. Nucleic Acid Degradation*

In 1962, Kessler and Engelberg[155] produced evidence for the loss of RNA from chloroplasts during leaf aging. A number of subsequent studies of chloroplasts in cell-free suspension have been reported, which show degradation of RNA and DNA during aging of this organelle.[75,156]

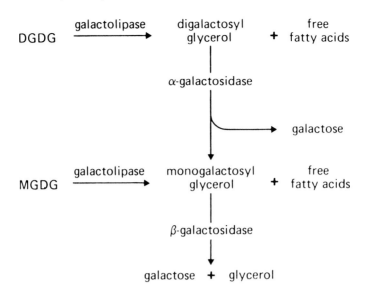

FIGURE 11. Hypothetical pathway for the degradation of MGDG and DGDG. (From Bhalla, P. L. and Dalling, M. J., *Plant Physiol.*, 76, 92, 1984. With permission.)

Purification of nucleic acids and ribonucleases from leaves and chloroplasts of wheat was described by Hadziyer et al.[157] They found the highest RNase activity of the subcellular particles to be localized in the chloroplasts, while ribosomes and mitochondria had very little activity associated with them. Particulate enzyme activity was shown to increase with shoot age, and to be further activated by cold conditions used during chloroplast isolation. Locy and Cherry[158] have described a chloroplast ribonuclease that degrades two of the four Tyr-t-RNA iso-accepting species from green soybean cotyledons. The susceptible Tyr-t-RNA species are also located in the chloroplast.

*4. Protein Degradation*
*a. Light Harvesting Chlorophyll a/b* **Protein**

The function of LHCP is to absorb light and subsequently transfer the excitation energy to either PS I or PS II.[259] LHCP contains about one third of the chlorophyll *a* and possibly all the chlorophyll *b* of the chloroplast (Section II.C.1). The direction of excitation energy transfer from LHCP to either Photosystem I or Photosystem II is controlled by reversible phosphorylation and dephosphorylation of the LHCP.[160,161] The polypeptides of LHCP are encoded by nuclear DNA[162-164] and are synthesized on cytoplasmic ribosomes as precursor proteins of molecular weight about 29.5 kdaltons. The precursor protein is larger than the mature apoprotein of LHCP by about 4 kdaltons.[163] Transport of the precursor protein from the cytoplasm to the stroma is ATP-dependent.[164] Processing of precursor polypeptides by a stroma protease is described in Chapter 4.

It would appear that the synthesis of LHCP and its assembly into the thylakoid membrane is under complex control. Bennett[159] has summarized the essential elements of this regulation and has identified four major mechanisms. First, phytochrome is able to influence synthesis of the necessary m-RNA. Second, degradation of the m-RNA is important. Third, light is essential for continued chlorophyll synthesis. Finally, the LHCP is subject to rapid degradation if the plants are kept in the dark; the apoprotein of the LHCP is also rapidly degraded if the assembled LHCP is not incorporated into the thylakoid membrane, or if there is insufficient chlorophyll for assembly of LHCP. Bennett concluded that the protease responsible for degradation of the LHCP is associated with the thylakoids. Lichtenthaler et al.[165] and Slovin and Tobin[166] also found the LHCP to be very unstable ($T_{1/2}$ about 10 hr) if plants were transferred to darkness. In addition, Slovin and Tobin describe an additional control of the level of LHCP. They suggest that light may somehow regulate synthesis of the LHCP apoprotein at the level of translation.

### b. $Q_B$-Protein (Peak D, Photogene 32)

In 1974, Eaglesham and Ellis[167] reported on the light-driven synthesis by intact pea chloroplasts of proteins that represented a quantitatively minor component of the thylakoids. The most prominently labeled protein had a molecular weight of about 32 kdaltons and was assigned the name Peak D. This protein is encoded by chloroplast DNA and is synthesized on chloroplast ribosomes as a 34.5 kdaltons precursor.[168] The relation between the low level of the protein in the chloroplast and the very high rate of its synthesis suggested that this protein was turned over at a rapid rate.[144] Peak D has now been identified as the primary binding site of the triazine herbicides.[169-171] The term $Q_B$ protein is now the accepted nomenclature for the 32 kdaltons protein (Peak D), which functions as the apoprotein of the bound quinone of electron transport.[170,171]

In addition to its important role in both electron transport and as a target for the triazine herbicides, the $Q_B$-protein is interesting on two other counts. First, Mattoo et al.[172] have shown that the $Q_B$-protein is stable in the dark and unstable in the light. Degradation was inhibited in the light by inhibitors of PS II electron transport, e.g., diuron, atrazine; ATP or photophosphorylation were apparently not involved. Second, Ohad et al.[171] have observed selective enhancement of the degradation of the $Q_B$-protein to be a primary step in the photoinhibition response of *Chlamydomonas* cells. Susceptibility of this protein to degradation was postulated to be due to the reaction of excess quinone anions with oxygen to produce an oxygen radical within the $Q_B$-protein.[170] This radical damaged the $Q_B$-protein in some manner, thereby increasing its susceptibility to degradation.

It is tempting to speculate that the high rate of degradation of the $Q_B$-protein is due to its participation in electron transport.[170,171] That is, the $Q_B$-protein is targeted by the reaction of oxygen or other radicals and as a consequence is "recognized" by the thylakoid protease responsible for its degradation. Under nonphotoinhibitory conditions, the production of targeted $Q_B$-proteins is less than the rate of synthesis of $Q_B$-protein. However, under photoinhibitory conditions, the entire plastoquinone pool is reduced and the rate of reaction of the $Q_B$-anion with oxygen is increased. Eventually all the $Q_B$-protein will become targeted, and, in these circumstances, the rate of $Q_B$-protein degradation far exceeds its rate of synthesis. As a consequence, it disappears from the thylakoid membrane and in this way affords the photosynthetic apparatus some short-term relief of the supra-optimal light intensity.

### c. NADPH 3-Protochlorophyllide Oxidoreductase

During the light-induced transformation of etioplasts into chloroplasts, the internal membranes are reorganized, some components of the membrane and chlorophyll are synthesized, and ultimately, the capacity for photosynthesis develops.[173] One important step in this phenomenon is the photoconversion of protochlorophyllide to chlorophyllide, catalyzed by the enzyme NADPH-protochlorophyllide oxidoreductase.

Within the first 5 min of exposing etiolated barley leaves to continuous white light, Santel and Apel[174] observed a 90% decline in the in vitro enzymatic activity of NADPH-protochlorophyllide oxidoreductase and more than 60% loss of the enzyme protein. Hamp and De Filippis[175] have described the activity of the two proteolytic enzymes associated with etioplasts of oat leaves and suggested that these enzymes may have a role in the light-induced de-differentiation of the prolamellar body structure. Plastid preparations from etiolated barley plants also contain proteolytic activity that will readily degrade hemoglobin, bovine serum albumin, and some proteins of the prolamellar body, including NADPH-protochlorophyllide oxidoreductase.[175] However, these authors could find no evidence to support a role for this proteolytic enzyme in the disappearance of NADPH-protochlorophyllide oxidoreductase. Recently a cell-free membrane preparation from etiolated barley leaves has been developed, which shows a light-dependent loss of NADPH-protochlorophyllide oxidoreductase protein.[177] Protein loss was temperature dependent and was independent of pH in the range 6 to 8.6. Protochlorophyllide and NADPH both appear to restrict degradation of the enzyme.

### d. Ribulose 1,5-Bisphosphate Carboxylase

RuBPCase turn-over in fully expanded leaves of *Zea mays* has been demonstrated unequivocably by Simpson et al.[178] The half-life was found to be between 7.8 and 6.5 days, depending on the method used. Concurrent synthesis and degradation of RuBPCase has also been demonstrated in the fully expanded 12th leaf of *Oryza sativa* L.[179] During senescence, the rate of degradation of RuBPCase far exceeded its rate of synthesis.

During assembly of the eight small and eight large subunits of RuBPCase (Section II.D.4), Schmidt and Mishkind[180] observed that SSU was rapidly degraded if it was not assembled into RuBPCase. They were unable to isolate the proteolytic enzyme responsible for this degradation and concluded that the protease was either very labile, or required unusual conditions for its extraction and in vitro assay. The protease is apparently synthesized on cytoplasmic ribosomes, since mutant cells of *Chlamydomonas*, which lack chloroplast ribosomes, still contain the enzyme.[180]

## IV. ROLE OF PROTEOLYTIC ENZYMES

Most of the following is largely speculative. There are no reports as yet that clearly implicate that any proteolytic enzyme has a specific or particular role in chloroplast senescence. There is, however, little doubt that chloroplasts contain a wide spectrum of hydrolytic enzymes, including proteolytic enzymes. From the evidence reviewed in this chapter, it would appear that the enzymes responsible for the degradative aspects of chloroplast senescence operate from within the chloroplast. In short, we dismiss the concept that chloroplast senescence is achieved through some pinocytotic relation with the vacuole.

### A. Influence of Nuclear and Chloroplast Genomes

While the specific enzymes have yet to be identified, there is reasonable circumstantial evidence to support the idea that some of the chloroplast hydrolytic enzymes may be coded for by nuclear DNA and synthesized on cytoplasmic ribosomes. This view has risen mainly from studies of the effects that protein synthesis inhibitors have on the initiation and expression of senescence in detached leaf segments. The earliest reports were those of Tavares et al.[181] and Knypl.[182] Since that time, the inhibitors have become more specific, with fewer side effects. Martin and Thimann[34] found that cycloheximide, an inhibitor of protein synthesis on 80S ribosomes (cytoplasm), retarded chlorophyll and protein degradation in detached leaves of oats and prevented the appearance of two proteolytic enzymes. Cycloheximide also prevents the appearance of RNase activity in excised leaves of *Lolium temulentum*; chloramphenicol, an inhibitor of protein synthesis on 70S ribosomes (chloroplast), did not retard senescence of detached leaves.[183] The D-isomer of the protein synthesis inhibitor (2-(4-methyl-2,6-dinitroanilino)-*N*-methylpropionamide (MDMP), which inhibits protein synthesis on 80S ribosomes, reduced the rate of $^{14}$C-leucine incorporation by detached leaf segments of *Festuca pratensis* by 95% compared with a control treatment.[184] D-MDMP retarded the loss of chlorophyll, RuBPCase protein, phosphoglycerokinase activity, and prevented any increase in RNase activity. The L-isomer of MDMP was without effect, and leaf segments treated with this compound senesced in a manner identical to the control.

From these and other experiments,[185-187] it is tempting to conclude that the synthesis of novel proteins is an integral part of senescence. However, while the data do not preclude such a possibility, they only offer strong evidence that continued protein synthesis is required for senescence to continue.

There are two additional lines of evidence that support the view that nuclear DNA exerts a significant influence on chloroplast senescence. First, Thomas and Stoddart[188] report that the nonyellowing mutant of *F. pratensis* (Section II.C.1) displays classical Mendelian inheritance, and therefore the gene(s) that control the expression of this character must be encoded by nuclear DNA. Second, in what is a most elegant demonstration of the influence that the nucleus has on the chloroplast, Yoshida[189] prepared enucleated cell halves of *Elodea densa* Casp. by inducing plasmolysis in a 0.2 $M$ solution of $CaCl_2$. Chloroplasts in the enucleated cell half did not senesce, while those chloroplasts in the cell half containing the nucleus senesced normally. In some cases, the plasmolysis treatment was incomplete and failed to produce distinct and separate cell halves; instead the cell halves were joined by a "plasma bridge". In these cases, senescence of chloroplasts in the enucleated cell half was similar to the nucleated cell half. Yoshida concluded that the "nuclear remote effect" could be transmitted through the cytoplasm to the chloroplast.

## B. Compartmentation of Proteolysis

Any speculation upon the hydrolytic events that might be associated with chloroplast senescence must include the studies of the nonyellowing mutant of *F. pratensis*.[81,82,138,139] These studies clearly indicate that senescence proceeds in two compartments within the chloroplast — the stroma and thylakoids. Processes occurring in the two compartments, in the mutant at least, are somewhat independent of one another. Therefore, we observe a normal pattern of breakdown of RuBPCase and the extrinsic proteins of the thylakoids. However, the intrinsic proteins, especially the chlorophyll:protein complexes, are not readily degraded (Section III.A.1). The mutation must, therefore, be directed towards the thylakoids.

Further speculation at this point may provide a useful framework for further studies.

### 1. Degradation of Thylakoid Proteins
#### a. Speculation No. 1

The nonyellowing mutant of *F. pratensis* may be of a gene(s) that directs in some manner the activity of a proteolytic enzyme in the hydrophobic environment of the thylakoid. Thomas[82] resolved the acid peptide hydrolase activity present in mature nonsenescent leaf tissue of *F. pratensis* into one major and three minor components, by ion exchange chromatography. The pattern of proteolytic enzymes in the wild-type was essentially identical to the nonyellowing mutant. In addition, the pH-response of enzyme activity directed towards various $^{14}C$-labeled protein fractions from the leaf were identical for both genotypes. Thomas has concluded that it is the inaccessibility of the thylakoid proteins in the mutant that determines their resistance to degradation. Although this conclusion seems reasonable, it fails to consider the possibility that the thylakoid proteins might be degraded by proteolytic enzymes associated with the thylakoid. The mutant may lack this enzyme. Furthermore, the use of crude extract as a basis for comparison is of limited value because nearly all the proteolytic activity detected will have originated from the vacuole (Section III.A).

Nettleton et al.[190] have recently shown that the major portion of the chloroplast proteolytic activity (RuBPCase as in vitro substrate) is associated with the thylakoids. At least part of this activity is associated with the surface of the thylakoid, but another enzyme with similar in vitro properties is more tightly associated or embedded in the membrane. Until the full extent of the diversity of the thylakoid proteolytic enzymes is known, our interpretation of the basis of nonyellowing in the *F. pratensis* mutant is restricted.

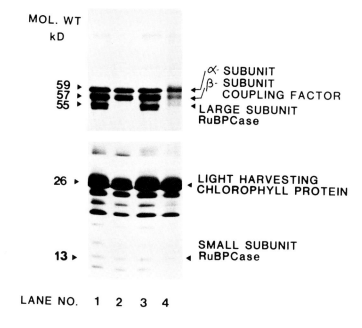

FIGURE 12. Degradation of the large subunit of RuBPCase and α and β subunits of Coupling Factor by proteolytic activity associated with thylakoids isolated from chloroplasts of the wheat primary leaf. Lanes 1 and 3 are zero time samples: lanes 2 and 4 assayed for 15 min in the absence and presence of 0.2% SDS, respectively. (From Nettleton, A. M., Bhalla, P. L., and Dalling, M. J., *J. Plant Physiol.*, 119, 35, 1985. With permission.)

### b. Speculation No. 2

Failure to degrade the intrinsic proteins of the thylakoids may not be due to a deficiency of proteolytic activity, but would be the consequence of the inability of the mutant to hydrolyze the thylakoid lipids. The thylakoid membrane environment is very complex due to the existence of both hydrophobic and hydrophilic domains and the possibility that many proteins have limited movement within the membrane. Susceptibility to degradation could be modulated through either the substrate or its respective proteolytic enzyme. In an intact, functioning membrane, access to the substrate may be restricted and consequently proteolysis is restricted. Dissolution, or some form of relaxation of the membrane, may be the mechanism that brings enzyme and substrate together.

In wheat thylakoids, severe dissolution of the thylakoids with the detergent SDS results in degradation of several abundant thylakoid proteins, e.g., α and β subunits of Coupling Factor and the apoprotein of the LHCP, by proteolytic activity associated with the thylakoids (Figure 12). Without detergent there is no apparent degradation. Detergent activation of chloroplast proteolytic activity has also been observed in barley[190] and *Pisum sativum* L.[191] These observations support the hypothesis that the integrity of the thylakoids is partially responsible for controlling the susceptibility to degradation of its constituent proteins.

Degradation of the thylakoid lipids is primarily the responsibility of galactolipase acting in concert with several other enzymes (Figure 11). A mutation affecting galactolipase activity could conceivably lead to an inability on the part of the mutant to degrade the thylakoid galactolipids. As a consequence, those proteins protected by either the steric rigidity or hydrophobic environment of the membrane would not be degraded. Unfortunately, we know little about the enzymatic basis for degradation of the thylakoid lipids and even less about the modulation or initiation of this phenomenon. In wheat leaves, galactolipase is firmly associated with the thylakoid membrane, and although very unstable, the enzyme has high in vitro activity.[193] β-Galactosidase is apparently a stromal enzyme.[154]

While crude extracts from the *Festuca* mutant apparently have acyl-hydrolase activity identical to the wild-type,[194] the possibility remains that a thylakoid-bound galactolipase would not be detected by this assay. Therefore, in much the same way as for the thylakoid protease activity, insufficient knowledge of the basic biochemistry involved finally puts an end to further speculation on lipolysis of the thylakoids.

*2. Degradation of Stromal Proteins*

Waters et al.[129] found that approximately half of the aminopeptidase activity in wheat mesophyll cells was associated with the stroma. While this enzyme had high exopeptidase activity against artificial substrates, some dipeptides and larger peptides, it had a very limited ability to degrade the main stromal protein RuBPCase. During senescence of the flag leaf of wheat, the aminopeptidase activity declined in parallel with leaf soluble protein and RuBPCase protein.[195]

ATP has recently been shown to initiate or stimulate the degradation of some stromal and thylakoid proteins, especially those of low-molecular-weight.[196,197] Malek et al.[197] suggest that this enzyme may be a means of degrading newly synthesized protein that is either aberrant or fails to assemble into a more complex state. In view of the dependence of the enzyme (or substrate) on ATP and the fact that photophosphorylation declines with senescence (Section II.D.3), it is difficult to envisage a role for this enzyme in senescence.

**C. Initiation of Proteolysis**

To date there are only three lines of evidence upon which to base any proposals for the initiation of proteolysis.

*1. Transcription of Nuclear DNA*

We have already reviewed the evidence (Section IV.A) that strongly supports the idea that continued protein synthesis, especially on cytoplasmic ribosomes, is an essential element of senescence. Unfortunately, there is little more to add to these findings at this stage. With the exciting developments being made in the field of molecular biology, however, it should not be long before the new techniques are being directed towards the senescence problem. Senescence in the molecular biological context is simply the outcome of gene expression. What causes the genes to be expressed is clearly of tremendous importance, not only to natural senescence, but also to senescence induced as a consequence of stress. Therefore, when one considers the problem of stress tolerance, especially in a plant breeding program, consideration is probably being made of the variation in the threshold level of some factor that allows for the expression of these genes.

*2. Targeting of Protein Substrates*

The hypothesis that proteins can be "targeted" makes the assumption that the necessary proteolytic enzymes (and other hydrolases) are already present in the chloroplast. With little modification, this hypothesis could accommodate the "Transcription of Nuclear DNA" hypothesis. Proteins are degraded after they have been "targeted" in some manner. In this review, we have considered two such mechanisms, one involving free radical damage of the $Q_B$-protein (Section III.B.4.b) and the other ATP (Section IV.B.2). Feller has also suggested that ATP and pyridine nucleotides may be able to protect some enzymes from proteolytic inactivation[198] (see Figure 4; Chapter 3).

*3. Activation by Unsaturated Fatty Acids*

The hypothesis that unsaturated fatty acids can activate proteolysis is directed specifically at the thylakoid-localized enzymes; it is not likely these enzymes will be responsible for degradation of stromal protein (Section IV.B). The precedent for this hypothesis is the recent work on the catabolism of chlorophyll.[98] This work indicates that unsaturated, long chain fatty acids, e.g., linolenic acid, the dominant fatty acid of the thylakoid lipids, can activate a chlorophyll oxidase. According to this hypothesis, degradation of the thylakoid lipids, with the concomitant release of their polyunsaturated fatty acids, is a primary event of senescence (Section III.B.2). This sets in motion a cascade, which includes proteolysis and chlorophyll degradation. This hypothesis is appealing because it unites the various degradative events responsible for thylakoid senescence. One major weakness of the hypothesis, however, is that the thylakoid lipids are presumed to be in a constant state of turn-over, regardless of the commencement of senescence. Perhaps it is the reactions that dispose of the released fatty acids that are important; if they were to decline for any reason, the pool of free fatty acids would increase, thereby initiating the cascade of events associated with senescence.

## REFERENCES

1. **Mohl,** cited by W. Pfeffer in his text *The Physiology of Plants, A Treatise Upon the Metabolism and Sources of Energy in Plants,* Vol. 1, 2nd ed., translated from German by A. J. Ewart, Clarendon Press, Oxford, 1899, 335.
2. **Schimper, A. F. W.**, Untersuchungen uber die chlorophyllkorper und die ihnen homologen gebilde, *Jahrb. Wiss. Bot.,* 16, 1, 1885.
3. **Whatley, J. M.**, A suggested cycle of plastid developmental interrelationships, *New Phytol.,* 80, 489, 1978.
4. **Thomson, W. W. and Whatley, J. M.**, Development of nongreen plastids, *Annu. Rev. Plant Physiol.,* 31, 375, 1980.
5. **Frey-Wyssling, A. and Kreutzen, E.**, Die Submikoskopische Entwicklung der Chromoplasten in den Bluten von *Ranunculus repens* L., *Planta,* 5, 104, 1958.
6. **Lewis, F. J. and Tuttle, G. M.**, Osmotic properties of some plant cells at low temperatures, *Am. J. Bot.,* 34, 405, 1920.
7. **Zacharova, T. M.**, Uber den Gasstoffwechsel der Nadelholzpflanzen in Winter, *Planta,* 8, 68, 1929.
8. **Schmidt, E.**, Baumgrenzenstudien am Feldberg im Schwarzwald, *Tharandt. Forstl. Jahrb.,* 87, 1, 1936.
9. **Chabot, J. F. and Chabot, B. F.**, Development and seasonal patterns of mesophyll ultrastructure in *Albies balsamea, Can. J. Bot.,* 53, 295, 1975.
10. **Senser, M., Schotz, F., and Beck, E.**, Seasonal changes in structure and function of spruce chloroplasts, *Planta,* 126, 1, 1975.
11. **Harris, W. H.**, Ultrastructural observations on the mesophyll cells of pine leaves, *Can. J. Bot.,* 49, 1107, 1971.
12. **Dodge, J. D.**, Changes in chloroplast fine structure during the autumnal senescence of *Betula* leaves, *Ann. Bot.,* 34, 817, 1970.
13. **Cunninghame, M. E., Hillman, J. R., and Bowes, B. G.**, Ultrastructural changes in mesophyll cells of *Larix decidus* × *kaempferi* during leaf maturation and senescence, *Flora,* 172, 161, 1982.
14. **Cunninghame, M. E., Bowes, B. G., and Hillman, J. R.**, An ultrastructural study of foliar senescence in *Taxus baccata* L., *Ann. Bot.,* 43, 527, 1979.
15. **Wardley, T. M., Bhalla, P. L., and Dalling, M. J.**, Changes in the number and composition of chloroplasts during senescence of mesophyll cells of attached and detached primary leaves of wheat (*Triticum aestivum* L.), *Plant Physiol.,* 74, 421, 1984.
16. **Wittenbach, V. A., Lin, W., and Herbert, R. R.**, Vacuolar localization of protease and degradation of chloroplasts in mesophyll protoplasts from senescing primary wheat leaves, *Plant Physiol.,* 69, 98, 1982.
17. **Lamppa, G. K., Elliot, L. V., and Bendich, A. J.**, Changes in chloroplast number during pea leaf development. An analysis of a protoplast population, *Planta,* 148, 437, 1980.
18. **Martinoia, E., Heck, U., Dalling, M. J., and Matile, Ph.**, Changes in chloroplast number and chloroplast constituents in senescing barley leaves, *Biochem. Physiol. Pflanzen.,* 178, 147, 1983.
19. **Ikeda, T. and Ueda, R.**, Light and electron-microscope studies on the senescence of chloroplasts in *Elodea* leaf cells, *Bot. Mag.,* 77, 336, 1964.
20. **Dennis, D. T., Stubbs, M., and Coultate, T. P.**, The inhibition of Brussels Sprout leaf senescence, *Can. J. Bot.,* 45, 1019, 1967.

21. **Ljubesic, N.,** Feinbau der Chloroplasten Wahrend der Vergilbung und Wiederergrunung der Blatter, *Protoplasma,* 66, 368, 1968.
22. **Mittelheuser, C. J. and Van Steveninck, R. F. M.,** The ultrastructure of wheat leaves. II. The effects of kinetin and ABA on detached leaves incubated in the light, *Protoplasma,* 73, 253, 1971.
23. **Barton, R.,** Fine structure of mesophyll cells in senescing leaves of *Phaseolus, Planta,* 71, 314, 1966.
24. **Sveshnikova, I. N., Kulaeva, O. N., and Bolyakina, Yu. P.,** Obrazovanie lamell i gran v khloroplastakh zheltyth list'ev pod deistviem 6-benzilaminopurina. (Lamellae and grana formation in chloroplasts of yellow leaves induced by 6-benzylaminopurine), *Fiziol. Rast.,* 13, 769, 1966.
25. **Butler, R. D.,** The fine structure of senescing cotyledons of cucumber, *J. Exp. Bot.,* 18, 535, 1967.
26. **Peoples, M. B., Beilharz, V. C., Waters, S. P., Simpson, R. J., and Dalling, M. J.,** Nitrogen redistribution during grain growth in wheat (*Triticum aestivum* L.). II. Chloroplast senescence and the degradation of ribulose-1,5-bisphosphate carboxylase, *Planta,* 149, 241, 1980.
27. **Wittenbach, V. A., Ackerson, R. C., Giaquinta, R. T., and Herbert, R. R.,** Changes in photosynthesis, ribulose bisphosphate carboxylase, proteolytic activity and ultrastructure of soybean leaves during senescence, *Crop Sci.,* 20, 225, 1980.
28. **Hurkman, W. J.,** Ultrastructural changes of chloroplasts in attached and detached, aging primary wheat leaves, *Am. J. Bot.,* 66, 64, 1979.
29. **Mittelheuser, C. J. and Van Steveninck, R. F. M.,** The ultrastructure of wheat leaves. I. Changes due to natural senescence and the effects of kinetin and ABA on detached leaves incubated in the dark, *Protoplasma,* 73, 239, 1971.
30. **Lichtenthaler, H. K.,** Plastoglobuli and the fine structure of plastids, *Endeavour,* 27, 144, 1968.
31. **Lichtenthaler, H. K.,** Die plastoglobuli von spinat, ihre grosse, isoliuerung und lipchinonzusammensetzung, *Protoplasma,* 68, 65, 1969.
32. **Barr, R. and Arntzen, C. J.,** The occurrence of α-tocophenylquinone in higher plants and its relation to senescence, *Plant Physiol.,* 44, 591, 1969.
33. **Lichtenthaler, H. K. and Weinert, H.,** The correlation between lipoquinone accumulation and plastoglobuli formation in the chloroplasts of *Ficus elastica* Roxb., *Z. Naturforsch.,* 256, 619, 1970.
34. **Martin, C. and Thimann, K. V.,** The role of protein synthesis in the senescence of leaves. I. The formation of protease, *Plant Physiol.,* 49, 64, 1972.
35. **Wittenbach, V. A.,** Induced senescence of intact wheat seedlings and its reversibility, *Plant Physiol.,* 59, 1039, 1977.
36. **Wittenbach, V. A.,** Breakdown of ribulose bisphosphate carboxylase and change in proteolytic activity during dark-induced senescence of wheat seedlings, *Plant Physiol.,* 62, 604, 1978.
37. **Shaw, M. and Manocha, M. S.,** Fine structure in detached, senescing wheat leaves, *Can. J. Bot.,* 43, 747, 1965.
38. **Ragetli, H. W. J., Weintraub, M., and Lo, E.,** Degeneration of leaf cells resulting from starvation after excision. I. Electron microscopic observations, *Can. J. Bot.,* 48, 1913, 1970.
39. **Mlodzianowski, F. and Kwintkiewicz, M.,** The inhibition of Kohlrabi chloroplast degeneration by kinetin, *Protoplasma,* 76, 211, 1973.
40. **Colquhoun, A. J., Hillman, J. R., Crewe, C., and Bowes, B. G.,** An ultrastructural study of the effect of abscisic acid in senescence of leaves of radish (*Raphanus sativus* L.), *Protoplasma,* 84, 205, 1975.
41. **Cohen, A. S., Popovic, R. B., and Zalik, S.,** Effect of polyamines on chlorophyll and protein content, photochemical activity, and chloroplast ultrastructure of barley leaf discs during senescence, *Plant Physiol.,* 64, 717, 1979.
42. **Richmond, A. E. and Lang, A.,** Effect of kinetin on protein content and survival of detached Xanthium leaves, *Science,* 125, 650, 1957.
43. **Osborne, D. J.,** Effect of kinetin on protein and nucleic acid metabolism in Xanthium leaves during senescence, *Plant Physiol.,* 37, 595, 1962.
44. **Sugiura, M., Umemura, K., and Oota, Y.,** The effect of kinetin on protein level of tobacco leaf disks, *Physiol. Plant.,* 15, 457, 1962.
45. **Shaw, M., Bhattacharya, P. K., and Quick, W. A.,** Chlorophyll, protein, and nucleic acid levels in detached, senescing wheat leaves, *Can. J. Bot.,* 43, 739, 1965.
46. **Fletcher, R. A.,** Retardation of leaf senescence by benzyladenine in intact bean plants, *Planta,* 89, 1, 1969.
47. **Atkin, R. K. and Sahai Srivastava, B. I.,** Studies on protein synthesis by senescing and kinetin-treated barley leaves, *Physiol. Plant.,* 23, 304, 1970.
48. **Tung, H. F. and Brady, C. J.,** Kinetin treatment and protein synthesis in detached wheat leaves, in *Proc. 7th Int. Conf. Plant Growth Substances,* 1972, 589.
49. **Mlodzianowski, F. and Ponitka, A.,** Ultrastructural changes of chloroplasts in detached parsley leaves yellowing in darkness and the influence of kinetin on that process, *Z. Pflanzenphysiol.,* 69, 13, 1973.
50. **Peterson, L. W. and Huffaker, R. C.,** Loss of ribulose 1,5-diphosphate carboxylase and increase in proteolytic activity during senescence of detached primary barley leaves, *Plant Physiol.,* 55, 1009, 1975.

51. **Biswal, U. C. and Mohanty, P.**, Changes in the ability of photophosphorylation and activities of surface-bound adenosine triphosphatase and ribulose diphosphate carboxylase of chloroplasts isolated from the barley leaves senescing in darkness, *Physiol. Plant.*, 44, 127, 1978.
52. **Wrischer, M.**, Ultrastructural changes in plastids of detached spinach leaves, *Z. Pflanzenphysiol.*, 86, 95, 1978.
53. **Misra, A. N. and Biswal, U. C.**, Effect of phytohormones on chlorophyll degradation during aging of chloroplasts *in vivo* and *in vitro*, *Protoplasma*, 105, 1, 1980.
54. **Fukuda, K. and Toyama, S.**, Electron microscope studies on the morphogenesis of plastids. XI. Ultrastructural changes of the chloroplasts in tomato leaves treated with ethylene and kinetin, *Cytoloqia*, 47, 725, 1982.
55. **Sirastava, B. J. S. and Arglebe, C.**, Effect of kinetin on ribosomes of excised barley leaves, *Physiol. Plant.*, 21, 851, 1968.
56. **Brady, C. J., Patterson, B. D., Tung, H. F., and Smillie, R. M.**, Protein and RNA synthesis during aging of chloroplasts in wheat leaves, in *Autonomy and Biogenesis of Mitochondria and Chloroplasts*, Boardman, N. K., Linnane, A. W., and Smillie, R. M., Eds., North-Holland, Amsterdam, 1971, 453.
57. **Naito, K., Ueda, K., and Tsuji, H.**, Differential effects of benzyladenine on the ultrastructure of chloroplasts in intact bean leaves according to their age, *Protoplasma*, 105, 293, 1981.
58. **Yang, S. F. and Hoffman, N. E.**, Ethylene biosynthesis and its regulation in higher plants, *Annu. Rev. Plant Physiol.*, 35, 155, 1984.
59. **Yang, S. F.**, Biosynthesis of ethylene and its regulation, in *Recent Advances in the Biochemistry of Fruit and Vegetables*, Friend, J. and Rhodes, M. J. C., Eds., Academic Press, London, 1981, 89.
60. **Toyama, S.**, Electron microscope studies on the morphogenesis of plastids. VII. Effects of streptomycin on the development of plastids in tomato cotyledon, *Bot. Mag. Tokyo*, 85, 89, 1972.
61. **Margulies, M. M.**, Effect of chloramphenicol on formation of chloroplast structure and protein during greening of etiolated leaves of *Phaseolus vulgaris*, *Plant Physiol.*, 41, 992, 1966.
62. **Endress, A. G. and Sjolund, R. D.**, Ultrastructural cytology of callus cultures of *Streptanthus tortuosus* as affected by temperature, *Am. J. Bot.*, 63, 1213, 1976.
63. **Schafers, H. A. and Feirabend, J.**, Ultrastructural differentiation of plastids and other organelles in rye leaves with a high-temperature-induced deficiency of plastid ribosomes, *Cytobiologie*, 14, 75, 1976.
64. **Possingham, J. V., Vesk, M., and Mercer, F. V.**, The fine structure of leaf cells of manganese-deficient spinach, *J. Ultrastruct. Res.*, 11, 68, 1964.
65. **Saikawa, M. and Ueda, R.**, Occurrence of phytoferritin particles during the development of plastids in variegated leaves of Japanese spindle trees, *Sci. Rep. Tokyo Kyoiku Daigaku*, 14B, 1, 1969.
66. **Toyama, S.**, Electron microscope studies on the morphogenesis of plastids. X. Ultrastructural changes of chloroplasts in Morning Glory leaves exposed to ethylene, *Am. J. Bot.*, 67, 625, 1980.
67. **Thorben, J. P.**, Chlorophyll-proteins: light-harvesting and reaction center components of plants, *Annu. Rev. Plant Physiol.*, 26, 127, 1975.
68. **Ryrie, I. J., Anderson, J. M., and Goodchild, D. J.**, The role of the light-harvesting chlorophyll a/b-protein complex in chloroplast membrane stacking. Cation-induced aggregation of reconstituted proteoliposomes, *Eur. J. Biochem.*, 107, 345, 1980.
69. **Anderson, J. M.**, P700 content and polypeptide profile of chlorophyll-protein complexes of spinach and barley thylakoids, *Biochim. Biophys. Acta*, 591, 113, 1980.
70. **Miller, K. R. and Staehelin, L. A.**, Analysis of the thylakoid outer surface. Coupling factor is limited to unstacked membrane regions, *J. Cell Biol.*, 68, 30, 1976.
71. **Andersson, B. and Anderson, J. M.**, Lateral heterogeneity in the distribution of chlorophyll-protein complexes of the thylakoid membranes of spinach chloroplasts, *Biochim. Biophys. Acta*, 593, 427, 1980.
72. **Sestak, Z.**, Photosynthetic characteristics during ontogenesis of leaves. II. Photosystems, components of electron transport chain, and photophosphorylation, *Photosynthetica*, 11, 449, 1977.
73. **Biswal, U. C., Bergfeld, R., and Kasemir, H.**, Phytochrome-mediated delay of plastid senescence in mustard cotyledons: changes in pigment contents and ultrastructure, *Planta*, 157, 85, 1983.
74. **Whitfield, D. M. and Rowan, K. S.**, Changes in the chlorophylls and carotenoids of *Nicotiana tabacum* during senescence, *Phytochemistry*, 13, 77, 1974.
75. **Panigrahi, P. K. and Biswal, U. C.**, Aging of chloroplasts *in vitro*. I. Quantitative analysis of the degradation of pigments, proteins and nucleic acids, *Plant Cell Physiol.*, 20, 775, 1979.
76. **Bricker, T. M. and Newman, D. W.**, Quantitative changes in the chloroplast thylakoid polypeptide complement during senescence, *Z. Pflanzenphysiol.*, 98, 339, 1980.
77. **Sestak, Z.**, Ratio of photosystem 1 and 2 particles in young and old leaves of spinach and radish, *Photosynthetica*, 3, 285, 1969.
78. **Sestak, Z., Zima, J., and Strnadova, H.**, Ontogenetic changes in the internal limitations to bean-leaf photosynthesis. II. Activities of photosystems 1 and 2 and non-cyclic photophosphorylation and their dependence on photon flux, *Photosynthetica*, 11, 282, 1977.

79. **Bricker, T. M. and Newman, D. W.**, Changes in the chlorophyll-proteins and electron transport activities of soybean (*Glycine max* L., cv. Wayne) cotyledon chloroplasts during senescence, *Photosynthetica*, 16, 239, 1982.
80. **Jenkins, G. I., Baker, N. R., and Woolhouse, H. W.**, Changes in chlorophyll content and organisation during senescence of the primary leaves of *Phaseolus vulgaris* L. in relation to photosynthetic electron transport, *J. Exp. Bot.*, 32, 1009, 1981.
81. **Thomas, H.**, Leaf senescence in a non-yellowing mutant of *Festuca pratensis*. I. Chloroplast membrane polypeptides, *Planta*, 154, 212, 1982.
82. **Thomas, H.**, Leaf senescence in a non-yellowing mutant of *Festuca pratensis*. II. Proteolytic degradation of thylakoid and stroma polypeptides, *Planta*, 154, 219, 1982.
83. **Douce, R. and Joyard, J.**, Plant galactolipids, in *The Biochemistry of Plants, a Comprehensive Treatise*, Vol. 4, Academic Press, New York, 1980, 321.
84. **Draper, S. R.**, Lipid changes in senescing cucumber cotyledons, *Phytochemistry*, 8, 1641, 1969.
85. **Ferguson, C. H. R. and Simon, E. W.**, Membrane lipids in senescing green tissues, *J. Exp. Bot.*, 24, 307, 1973.
86. **Fong, F. and Heath, R. L.**, Age dependent changes in phospholipids and galactolipids in primary bean leaves (*Phaseolus vulgaris*), *Phytochemistry*, 16, 215, 1977.
87. **Chia, L. S., Thompson, J. E., and Dumbroff, E. B.**, Simulation of the effects of leaf senescence on membranes by treatment with paraquat, *Plant Physiol.*, 67, 415, 1981.
88. **McKersie, B. D. and Thompson, J. E.**, Phase behaviour of chloroplast and microsomal membranes during leaf senescence, *Plant Physiol.*, 61, 639, 1978.
89. **Woolhouse, H. W. and Batt, T.**, The nature and regulation of senescence in plastids, *Perspect. Exp. Biol.*, 2, 163, 1976.
90. **Friedrich, J. W. and Huffaker, R. C.**, Photosynthesis, leaf resistances and ribulose-1,5-bisphosphate carboxylase degradation in senescing barley leaves, *Plant Physiol.*, 65, 1103, 1980.
91. **Hall, N. P., Keys, A. J., and Merret, M. J.**, Ribulose-1,5-diphosphate carboxylase protein during flag leaf senescence, *J. Exp. Bot.*, 29, 31, 1978.
92. **Camp, P. J., Huber, S. C., Burke, J. J., and Moreland, D. E.**, Biochemical changes that occur during senescence of wheat leaves. I. Basis for the reduction of photosynthesis, *Plant Physiol.*, 70, 1641, 1982.
93. **Makino, A., Mae, T., and Ohira, K.**, Photosynthesis and ribulose 1,5-bisphosphate carboxylase in rice leaves. Changes in photosynthesis and enzymes involved in carbon assimilation from leaf development through senescence, *Plant Physiol.*, 73, 1002, 1983.
94. **Camp, P. J., Huber, S. C., and Moreland, D. E.**, Changes in enzymes of sucrose metabolism and the activation of certain chloroplast enzymes during wheat leaf senescence, *J. Exp. Bot.*, 35, 659, 1984.
95. **Evans, J. R.**, Nitrogen and photosynthesis in the flag leaf of wheat (*Triticum aestivum* L.), *Plant Physiol.*, 72, 297, 1983.
96. **Sestak, Z.**, Photosynthetic characteristics during ontogenesis of leaves. II. Photosystems, components of electron transport chain, and photophosphorylations, *Photosynthetica*, 11, 449, 1977.
97. **Percival, M. P., Williams, W. P., Chapman, D., and Quinn, P. J.**, Loss of Hill activity in isolated chloroplasts is not directly related to free fatty acid release during aging, *Plant Sci. Lett.*, 19, 47, 1980.
98. **Martinoia, E., Dalling, M. J., and Matile, Ph.**, Catabolism of chlorophyll: demonstration of chloroplast-localized peroxidative and oxidative activities, *Z. Pflanzenphysiol.*, 107, 269, 1982.
99. **Katoh, S. and San Pietro, A.**, A comparative study of the inhibitory action on the oxygen-evolution system of various chemical and physical treatments of *Euglena* chloroplasts, *Arch. Biochem. Biophys.*, 128, 378, 1968.
100. **Cohen, W. S., Nathanson, B., White, J. E., and Brody, M.**, Fatty acids as model systems for the action of *Ricinus* leaf extract on higher plant chloroplasts and algae, *Arch. Biochem. Biophys.*, 135, 21, 1969.
101. **Siegenthaler, P. A.**, Aging of the photosynthetic apparatus. IV. Similarity between the effects of aging and unsaturated fatty acids on isolated chloroplasts as expressed by volume changes, *Biochim. Biophys. Acta*, 275, 182, 1972.
102. **Siegenthaler, P. A.**, Change in pH dependence and sequential inhibition of photosynthetic activity in chloroplasts by unsaturated fatty acids, *Biochim. Biophys. Acta*, 305, 153, 1973.
103. **Siegenthaler, P. A.**, Inhibition of photosystem II electron transport in chloroplasts by fatty acids and restoration of its activity by $Mn^{2+}$, *FEBS Lett.*, 39, 337, 1974.
104. **Siegenthaler, P. A. and Depery, F.**, Influence of unsaturated fatty acids in chloroplasts. Shift of the pH optimum of electron flow and relations to $\Delta pH$, thylakoid internal pH and proton uptake, *Eur. J. Biochem.*, 61, 573, 1976.
105. **Jenkins, G. I. and Woolhouse, H. W.**, Photosynthetic electron transport during senescence of the primary leaves of *Phaseolus vulgaris* L. II. The activity of photosystems one and two, and a note on the site of reduction of ferricyanide, *J. Exp. Bot.*, 32, 989, 1981.
106. **Jenkins, G. I. and Woolhouse, H. W.**, Photosynthetic electron transport during senescence of the primary leaves of *Phaseolus vulgaris* L. I. Non-cyclic electron transport, *J. Exp. Bot.*, 32, 467, 1981.

107. **Jenkins, G. I., Baker, N. R., Bradbury, M., and Woolhouse, H. W.**, Photosynthetic electron transport during senescence of the primary leaves of *Phaseolus vulgaris* L. III. Kinetics of chlorophyll fluorescence emission from intact leaves, *J. Exp. Bot.*, 32, 999, 1981.
108. **Holloway, P. J., Maclean, D. J., and Scott, K. J.**, Rate-limiting steps of electron transport in chloroplasts during ontogeny and senescence of barley, *Plant Physiol.*, 72, 795, 1983.
109. **Ben-David, H., Nelson, N., and Gepstein, S.**, Differential changes in the amount of protein complexes in the chloroplast membrane during senescence of oat and bean leaves, *Plant Physiol.*, 73, 507, 1983.
110. **Moon, K. E. and Thompson, E. O. P.**, Subunits from reduced and S-carboxymethylated ribulose diphosphate carboxylase (fraction 1 protein), *Aust. J. Biol. Sci.*, 22, 463, 1969.
111. **Rutner, A. C.**, Estimation of molecular weights of ribulose diphosphate carboxylase subunits, *Biochem. Biophys. Res. Commun.*, 39, 923, 1970.
112. **Yeoh, H. H., Stone, N. E., Creaser, E. H., and Watson, L.**, Isolation and characterization of wheat ribulose-1,5-diphosphate carboxylase, *Phytochemistry*, 18, 561, 1979.
113. **Jensen, R. G. and Bahr, J. T.**, Ribulose 1,5-bisphosphate carboxylase-oxygenase, *Annu. Rev. Plant Physiol.*, 28, 379, 1977.
114. **Blair, G. E. and Ellis, R. J.**, Protein synthesis in chloroplasts. I. Light-driven synthesis of the large subunit of Fraction 1 protein by isolated pea chloroplasts, *Biochim. Biophys. Acta*, 319, 223, 1973.
115. **Gray, J. C. and Kekwick, R. G. O.**, The synthesis of the small subunit of ribulose 1,5-bisphosphate carboxylase in the French Bean, *Phaseolus vulgaris*, *Eur. J. Biochem.*, 44, 491, 1974.
116. **Highfield, P. E. and Ellis, R. J.**, Synthesis and transport of the small subunit of chloroplast ribulose bisphosphate carboxylase, *Nature (London)*, 271, 420, 1978.
117. **Brady, C. J.**, A co-ordinated decline in the synthesis of subunits of ribulose bisphosphate carboxylase in aging wheat leaves. I. Analyses of isolated protein, subunits and ribosomes, *Aust. J. Plant Physiol.*, 8, 591, 1981.
118. **Kannangara, C. G. and Woolhouse, H. W.**, Changes in the enzyme activity of soluble protein fractions in the course of foliar senescence in *Perilla frutescens* (L.) Britt., *New Phytol.*, 67, 533, 1968.
119. **Callow, J. A., Callow, M. E., and Woolhouse, H. W.**, *In vitro* protein synthesis, ribosomal RNA synthesis and polyribosomes in senescing leaves of *Perilla.*, *Cell Differentiation*, 1, 79, 1972.
120. **Callow, J. A.**, Ribosomal RNA, fraction 1 protein synthesis, and ribulose diphosphate carboxylase activity in developing and senescing leaves of cucumber, *New Phytol.*, 73, 13, 1974.
121. **Dickman, D. C. and Gordon, J. C.**, Incorporation of $^{14}C$-photosynthate into protein during leaf development in young *Populus* plants, *Plant Physiol.*, 56, 23, 1975.
122. **Brady, C. J. and Tung, H. F.**, Rate of protein synthesis in senescing, detached wheat leaves, *Aust. J. Plant Physiol.*, 2, 163, 1975.
123. **Eilam, Y., Butler, R. D., and Simon, E. W.**, Ribosomes and polysomes in cucumber leaves during growth and senescence, *Plant Physiol.*, 47, 317, 1971.
124. **Brady, C. J. and Steele Scott, N.**, Chloroplast polyribosomes and synthesis of fraction I protein in the developing wheat leaf, *Aust. J. Plant Physiol.*, 4, 327, 1977.
125. **Spiers, J. and Brady, C. J.**, A co-ordinated decline in the synthesis of subunits of ribulose bisphosphate carboxylase in aging wheat leaves. II. Abundance of messenger RNA, *Aust. J. Plant Physiol.*, 8, 603, 1981.
126. **Ness, P. J. and Woolhouse, H. W.**, RNA synthesis in *Phaseolus* chloroplasts. II. Ribonucleic acid synthesis in chloroplasts from developing and senescing leaves, *J. Exp. Bot.*, 31, 235, 1980.
127. **Matile, Ph.**, *The Lytic Compartment of Plant Cells*, Springer-Verlag, New York, 1975, 18.
128. **Boller, T. and Kende, H.**, Hydrolytic enzymes in the central vacuole of plant cells, *Plant Physiol.*, 63, 1123, 1979.
129. **Waters, S. P., Noble, E. R., and Dalling, M. J.**, Intracellular localization of peptide hydrolases in wheat (*Triticum aestivum* L.) leaves, *Plant Physiol.*, 69, 575, 1982.
130. **Noble, E. R. and Dalling, M. J.**, Intracellular localization of acid peptide hydrolases and several other acid hydrolases in the leaf of pea (*Pisum sativum* L.), *Aust. J. Plant Physiol.*, 9, 353, 1982.
131. **Butcher, H. C., Wagner, G. J., and Siegelman, H. W.**, Localization of acid hydrolases in protoplasts — examination of the proposed lysosomal function of the mature vacuole, *Plant Physiol.*, 59, 1098, 1977.
132. **Buser, Ch. and Matile, Ph.**, Malic acid in vacuoles isolated from *Bryophyllum* leaf cells, *Z. Pflanzenphysiol.*, 82, 462, 1977.
133. **Baumgartner, B. and Matile, Ph.**, Immunocytochemical localization of acid ribonuclease in Morning Glory flower tissue, *Biochem. Physiol. Pflanz.*, 170, 279, 1976.
134. **Heck, U., Martinoia, E., and Matile, Ph.**, Subcellular localization of acid proteinase in barley mesophyll protoplasts, *Planta*, 151, 198, 1981.
135. **Lin, W. and Wittenbach, V. A.**, Subcellular localization of proteases in wheat and corn mesophyll protoplasts, *Plant Physiol.*, 67, 969, 1981.
136. **Wagner, G. J., Mulready, P., and Cuff, J.**, Vacuole/extravacuole distribution of soluble protease in *Hippeastrum* petal and *Triticum* leaf protoplasts, *Plant Physiol.*, 68, 1081, 1981.

137. **Thayer, S. S. and Huffaker, R. C.**, Vacuolar localization of endoproteinases EPI and EP2 in barley mesophyll cells, *Plant Physiol.*, 75, 70, 1984.
138. **Thomas, H. and Stoddart, J. L.**, Separation of chlorophyll degradation from other senescence processes in leaves of a mutant genotype of meadow fescue *(Festuca pratensis), Plant Physiol.*, 56, 438, 1975.
139. **Thomas, H.**, Ultrastructure, polypeptide composition and photochemical activity of chloroplasts during foliar senescence of a non-yellowing mutant genotype of *Festuca pratensis* Huds., *Planta*, 137, 53, 1977.
140. **Thomas, H.**, Proteolysis in senescing leaves, in Interactions Between Nitrogen and Growth Regulators in the Control of Plant Development, Monogr. 9, British Plant Growth Regulator Group, London, 1983, 45.
141. **Hetherington, S. E., Hallam, N. D., and Smillie, R. M.**, Ultrastructural and compositional changes in chloroplast thylakoids of leaves of *Borya nitida* during humidity-sensitive degreening, *Aust. J. Plant Physiol.*, 9, 601, 1982.
142. **Cooke, R. J., Roberts, K., and Davies, D. D.**, Model for stress-induced protein degradation in *Lemna minor, Plant Physiol.*, 66, 1119, 1980.
143. **Boulter, D., Ellis, R. J., and Yarwood, A.**, Biochemistry of protein synthesis in plants, *Biol. Rev.*, 47, 113, 1972.
144. **Ellis, R. J.**, Chloroplast proteins: synthesis, transport, and assembly, *Annu. Rev. Plant Physiol.*, 32, 111, 1981.
145. **Purvis, A. C. and Barmore, C. R.**, Involvement of ethylene in chlorophyll degradation in peel of citrus fruit, *Plant Physiol.*, 68, 854, 1981.
146. **Kuroki, M., Shioi, Y., and Sasa, T.**, Purification and properties of soluble chlorophyllase from tea leaf sprouts, *Plant Cell Physiol.*, 22, 717, 1981.
147. **Huff, A.**, Peroxidase-catalyzed oxidation of chlorophyll by hydrogen peroxide, *Phytochemistry*, 21, 261, 1982.
148. **Matile, Ph.**, Catabolism of chlorophyll: involvement of peroxidase?, *Z. Pflanzenphysiol.*, 99, 475, 1980.
149. **Sastry, P. S. and Kates, M.**, Hydrolysis of monogalactosyl and digalactosyl diglycerides by specific enzymes in runner-bean leaves, *Biochemistry*, 3, 1280, 1964.
150. **Hoshina, S., Kaji, T., and Nishida, K.**, Photoswelling and light-inactivation of isolated chloroplasts. I. Change in lipid content in light-aged chloroplasts, *Plant Cell Physiol.*, 16, 465, 1975.
151. **Constantopoulos, G. and Kenyon, C. N.**, Release of free fatty acids and loss of Hill activity of aging spinach chloroplasts, *Plant Physiol.*, 43, 531, 1968.
152. **Wright, A. J., and Fishwick, M. J.**, Lipid degradation during manufacture of black tea, *Phytochemistry*, 18, 1511, 1979.
153. **Anderson, M. N., McCarty, R. E., and Zimmer, E. A.**, The role of galactolipids in spinach chloroplast lamellar membranes. I. Partial purification of a bean leaf galactolipid lipase and its action on subchloroplast particles, *Plant Physiol.*, 53, 699, 1974.
154. **Bhalla, P. L. and Dalling, M. J.**, Characteristics of a β-galactosidase associated with the stroma of chloroplasts prepared from mesophyll protoplasts of the primary leaf of wheat, *Plant Physiol.*, 76, 92, 1984.
155. **Kessler, B. and Engelberg, N.**, Ribonucleic acid and ribonuclease activity in developing leaves, *Biochim. Biophys. Acta*, 55, 70, 1962.
156. **Misra, A. N. and Biswal, U. C.**, Changes in the content of plastid macromolecules during aging of attached and detached leaves, and of isolated chloroplasts of wheat seedlings, *Photosynthetica*, 16, 22, 1982.
157. **Hadziyer, D., Mehta, S. L., and Zalik, S.**, Nucleic acids and ribonucleases of wheat leaves and chloroplasts, *Can. J. Biochem.*, 47, 273, 1969.
158. **Locy, R. D. and Cherry, J. H.**, Evidence for a chloroplast specific tyrosyl tRNA degrading activity, *Biochem. Biophys. Res. Commun.*, 72, 15, 1976.
159. **Bennett, J.**, Biosynthesis of the light-harvesting chlorophyll a/b protein. Polypeptide turnover in darkness, *Eur. J. Biochem.*, 118, 61, 1981.
160. **Bennett, J., Steinback, K. E., and Arntzen, C. J.**, Chloroplast phosphoproteins: regulation of excitation energy transfer by phosphorylation of thylakoid membrane polypeptides, *Proc. Natl. Acad. Sci. U.S.A.*, 77, 5253, 1980.
161. **Allen, J. F., Bennet, J., Steinback, K. E., and Arntzen, C. J.**, Chloroplast protein phosphorylation couples plastoquinone redox state to distribution of excitation energy between photosystems, *Nature (London)*, 291, 25, 1981.
126. **Kung, S. D., Thornber, J. P., and Wildman, S. G.**, Nuclear DNA codes for the photosystem II chlorophyll-protein of chloroplast membranes, *FEBS Lett.*, 24, 185, 1972.
163. **Apel, K. and Kloppstech, K.**, The plastid membranes of barley (*Hordeum vulgare*). Light-induced appearance of mRNA coding for the apoprotein of the light-harvesting chlorophyll a/b protein, *Eur. J. Biochem.*, 85, 581, 1978.
164. **Grossman, A., Bartlett, S., and Chua, N-M.**, Energy-dependent uptake of cytoplasmically synthesized polypeptides by chloroplasts, *Nature (London)*, 285, 625, 1980.

165. **Lichtenthaler, H. K., Burkard, G., Kuhn, G., and Prenzel, U.**, Light-induced accumulation and stability of chlorophylls and chlorophyll-proteins during chloroplast development in radish seedlings, *Z. Naturforsch.*, 36c, 421, 1981.
166. **Slovin, J. P. and Tobin, E. M.**, Synthesis and turnover of the light-harvesting chlorophyll a/b-protein in *Lemna gibba* grown with intermittent red light: possible translational control, *Planta*, 154, 465, 1982.
167. **Eaglesham, A. R. J. and Ellis, R. J.**, Protein synthesis in chloroplasts II. Light-driven synthesis of membrane proteins by isolated pea chloroplasts, *Biochim. Biophys. Acta*, 335, 396, 1974.
168. **Grebanier, A. E., Coen, D. M., Rich, A., and Bogorad, L.**, Membrane proteins synthesised but not processed by isolated maize chloroplasts, *J. Cell. Biol.*, 78, 734, 1978.
169. **Steinback, K. E., Pfister, K., and Arntzen, C. J.**, Identification of the receptor site for triazine herbicides in chloroplast thylakoid membranes, in *Biochemical Responses Induced by Herbicides*, Moreland, D. E., St. John, J. B., and Hess, F. D., Eds., American Chemical Society, Washington, D.C., 1982, 37.
170. **Kyle, D. J., Ohad, I., and Arntzen, C. J.**, Membrane protein damage and repair: selective loss of a quinone-protein function in chloroplast membranes, *Proc. Natl. Acad. Sci. U.S.A.*, 81, 4070, 1984.
171. **Ohad, I., Kyle, D. J., and Arntzen, C. J.**, Membrane protein damage and repair: removal and replacement of inactivated 32-kilodalton polypeptides in chloroplast membranes, *J. Cell Biol.*, 99, 481, 1984.
172. **Mattoo, A. K., Hoffman-Falk, H., Marden, J. B., and Edelman, M.**, Regulation of protein metabolism: coupling of photosynthetic electron transport to *in vivo* degradation of the rapidly metabolized 32-kilodalton protein of the chloroplast membranes, *Proc. Natl. Acad. Sci. U.S.A.*, 81, 1380, 1984.
173. **Boardman, N. K., Anderson, J. M., and Goodchild, D. J.**, Chlorophyll-protein complexes and structure of mature and developing chloroplasts, *Curr. Top. Bioenerg.*, 8, 35, 1978.
174. **Santel, H.-J. and Apel, K.**, The protochlorophyllide holochrome of barley (*Hordeum vulgare* L.). The effect of light on the NADPH:protochlorophyllide oxidoreductase, *Eur. J. Biochem.*, 120, 95, 1981.
175. **Hampp, R. and De Filippis, L. F.**, Plastid protease activity and prolamellar body transformation during greening, *Plant Physiol.*, 65, 663, 1980.
176. **Dehesh, K. and Apel, K.**, The function of proteases during the light-dependent transformation of etioplasts to chloroplasts in barley (*Hordeum vulgare* L.), *Planta*, 157, 381, 1983.
177. **Hauser, I., Dehesh, K., and Apel, K.**, The proteolytic degradation *in vitro* of the NADPH-protochlorophyllide oxidoreductase of barley (*Hordeum vulgare* L.), *Arch. Biochem. Biophys.*, 228, 577, 1984.
178. **Simpson, E., Cooke, R. J., and Davies, D. D.**, Measurement of protein degradation in leaves of *Zea mays* using [$^3$H] acetic anhydride and tritiated water, *Plant Physiol.*, 67, 1214, 1981.
179. **Mae, T., Makino, A., and Ohira, K.**, Changes in the amounts of ribulose bisphosphate carboxylase synthesized and degraded during the life span of rice leaf (*Oryza sativa* L.), *Plant Cell Physiol.*, 24, 1079, 1983.
180. **Schmidt, G. W. and Mishkind, M. L.**, Rapid degradation of unassembled ribulose 1,5-bisphosphate carboxylase small subunits in chloroplasts, *Proc. Natl. Acad. Sci. U.S.A.*, 80, 2632, 1983.
181. **Tavares, J., Kende, H., and Berke, E.**, Action of benzyladenine on detached leaves of corn, *Annu. Rep. MSU/AEC Plant Res. Lab.*, p. 93, 1968.
182. **Knypl, J. S.**, Arrest of yellowing in senescing leaf discs of maize by growth retardants, coumarin and inhibitors of RNA and protein synthesis, *Biol. Plant.*, 12, 199, 1970.
183. **Thomas, H.**, Regulation of alanine aminotransferase in leaves of *Lolium temulentum* during senescence, *Z. Pflanzenphysiol. Bd.*, 74, 208, 1975.
184. **Thomas, H.**, Delayed senescence in leaves treated with the protein synthesis inhibitor MDMP, *Plant Sci. Lett.*, 6, 369, 1976.
185. **Yu, S. M. and Kao, C. H.**, Retardation of leaf senescence by inhibitors of RNA and protein synthesis, *Physiol. Plant.*, 52, 207, 1981.
186. **Pjon, C.-J.**, Effects of cycloheximide and light on leaf senescence in maize and hydrangea, *Plant Cell Physiol.*, 22, 847, 1981.
187. **Garcia, S., Martin, M., and Sabater, B.**, Protein synthesis by chloroplasts during the senescence of barley leaves, *Physiol. Plant.*, 57, 260, 1983.
188. **Thomas, H. and Stoddart, J. L.**, Leaf senescence, *Annu. Rev. Plant Physiol.*, 31, 83, 1980.
189. **Yoshida, Y.**, Nuclear control of chloroplast activity in *Elodea* leaf cells, *Protoplasma*, 54, 476, 1961.
190. **Nettleton, A. M., Bhalla, P. L., and Dalling, M. J.**, Characterization of peptide hydrolase activity associated with thylakoids of the primary leaves of wheat, *J. Plant Physiol.*, 119, 35, 1985.
191. **Dalling, M. J., Tang, A. B., and Huffaker, R. C.**, Evidence for the existence of peptide hydrolase activity associated with chloroplasts isolated from barley mesophyll protoplasts, *Z. Pflanzenphysiol.*, 111, 311, 1983.
192. **Dahlmelm, H. and Ficker, K.**, Investigation on the subcellular localization of proteolytic enzymes in *Pisum sativum* L. II. Proteolytic activity in chloroplasts, *Biochem. Physiol. Pflanzen.*, 177, 167, 1982.
193. **O'Sullivan, J.N. and Dalling, M. J.**, unpublished data.
194. **Harwood, J. L., Jones, A. V. H. M., and Thomas, H.**, Leaf senescence in a non-yellowing mutant of *Festuca pratensis*. III. Total acyl lipids of leaf tissue during senescence, *Planta*, 156, 152, 1982.

195. **Waters, S. F., Peoples, M. B., Simpson, R. J., and Dalling, M. J.,** Nitrogen redistribution during grain growth in wheat (*Triticum aestivum* L.). I. Peptide hydrolase activity and protein breakdown in flag leaf, glumes and stem, *Planta,* 148, 422, 1980.
196. **Liu, X.-Q. and Jagendorf, A. T.,** ATP-dependent proteolysis in pea chloroplasts, *FEBS Lett.,* 166, 248, 1984.
197. **Malek, L., Bogorad, L., Ayers, A., and Goldberg, A.,** Nearly synthesized proteins are degraded by an ATP-stimulated proteolytic process in isolated pea chloroplasts, *FEBS Lett.,* 166, 253, 1984.
198. **Streit, L. and Feller, U.,** Inactivation of N-assimilating enzymes and proteolytic activities in wheat leaf extracts: effect of pyridine nucleotides and of adenylates, *Experientia,* 38, 1176, 1982.

# INDEX

## A

Acalin, B, 25
Adenosine diphosphate (ADP), 61
Adenosine monophosphate (AMP), 61
Adenosine triphosphate (ATP), 61, 62
Adzuki bean, root nodules, protease activity, 115, 118
Agglutinin, *Ricinus communis*, 74
Alanine
  in cereal albumins, 4
  in cereal globulins, 4
  in cereal glutelins, 7
  in cereal prolamins, 6
Albumins(s), see also Storage proteins, 22
  amino acid composition
    in cereals, 4
    in dicots, 24
  of cereals, 3, 4
  of dicots, 24, 27
  low molecular weight, of dicots, 27
Aleurone, protein bodies, 2
Aleurone grains, 23
Alfalfa root nodules, 106—107
  nitrogen fixation, 105
  structure and function after shoot removal, 120, 121
Alga
  *Poterioochromonas*, 92—101
  *Valonia*, 92
Alkaline peptidases, of dicots, 31—32
Amino acids
  of Bowman-Birk soybean trypsin inhibitor, 24, 26
  of cereal albumins, 3, 4
  of cereal globulins, 3, 4
  of cereal glutelins, 6, 7
  of cereal prolamins, 5, 6
  of dicot storage proteins, 24
  of Kunitz soybean trypsin inhibitor, 24
  of lectins, 24
  of legumins, 23, 24
  signal sequence, 70
  of vicilin, 24
Aminopeptidases
  leucine, 31
  in senescing leaves, 51—52
  soybean nodules, 115
Ammonium, root nodule function and, 105
*Ananas*, see Pineapple
Andropoganeae, see Maize; Sorghum
1-Anilino-8-naphthalenesulfonate (ANS), in pumpkin endopeptidase assay, 29
Anson's assay, for plant proteolytic activity, 28
Apple tree, leaf senescence, proteases in, 55
Arachin, 23
*Arachis hypogeae*, see also Peanut, 21
Arginine
  in cereal albumins, 4
  in cereal globulins, 4
  in cereal glutelins, 7
  in cereal prolamins, 6
Arylamidases
  of dicots, 31
  of pumpkin, 35
  of vetch seeds, 36
Asparagine, in storage protein processing, 72, 73
Asparagus, green, proteolytic activity, 55
Aspartic acid
  in cereal albumins, 4
  in cereal globulins, 4
  in cereal glutelins, 7
  in cereal prolamins, 6
Aspartic proteinase, of dry soybean, 29
Azocasein degradation, 56
Azocasein protease activity
  in adzuki bean nodules, 115, 118
  in alfalfa nodule, 121
  in soybean nodules, 115
Azocollase A, 54, 56
Azocollase B, 54, 56

## B

Bacteroid-containing cells, 104, 106—109
Barley (*Hordeum*)
  albumins, 3, 4
  globulins, 4
  glutelins, 6, 7
  leaf
    electron transport during senescence, 135
    proteinase, 83, 84, 86, 87
  prolamins, 3, 5, 6
  protein body formation in, 2
  proteolytic enzymes in, 10—12
  storage protein hydrolysis in, 7—8
    proteases and, 10—12
BBSTI, see Bowman-Birk soybean trypsin inhibitor
Bean, see also specific types
  broad (*Vicia*), 23
  castor (*Ricinus*)
    carboxypeptidases, 30
    storage proteins, 27
  common (*Phaseolus*)
    storage proteins, 21
    vicilin, 37
  great northern, endopeptidase, 29
  leaves, endopeptidase activity, 56—57
*Betula* leaves, autumnal senescence, chloroplast shape and volume changes, 127, 130
Birdsfoot trefoil
  defoliation, 105
  root nodule, 110
    senescence, 112—113
*Borya nitida* Labill, 138
Bovine serum albumin, as nitrate reductase protector, 84

Bowman-Birk soybean trypsin inhibitor (BBSTI), 26, 27
  amino acid composition, 24, 26
  degradation, 38—39
  properties, 26
Broad bean, see *Vicia faba*
Buoyant density, of senescing chloroplasts, 127, 130

## C

Calcium, generation of IFP-synthase-activating proteinase, 97—98, 100
Calmodulin, in generation of IFP-synthase-activating proteinase from membranes, 97—98
*Canavalia ensiformis*, 25, 30
*Cannabis sativa*, see also Hemp, 29
Carbohydrates, 20
  in vicilins, 71
Carboxypeptidase
  in barley, 11, 12
  in dicots, 30—31
  inhibitors of, 11
  in maize, 12—13
  in rice, 13
  in senescing leaves, 52—53
  in sorghum, 13
  in soybean nodules, 115
  in storage protein hydrolysis, 10—13
  in wheat, 10—11
*Carica papaya*, 55
Caricaceae, 55
Carotenoids, in chloroplast senescence, 131
Casein, as nitrate reductase protector, 84
Casein substrate, 53
Caseolytic activity, in wheat leaf, 58
Castor bean (*Ricinus*), 27, 30
Cell, standard volume of, 100
Cell shrinkage, and IFP-synthase activation, 98
Cell volume, regulation of, in *Poterioochromonas*, 92—93
Cereals
  albumins, 3, 4
  globulins, 3, 4
  leaf senescence, endopeptidase during, 53—54, 56
  prolamins, 3—5, 6
  protein body formation in endosperm, 2—3
  storage proteins, 72
    hydrolysis, 6—10
Chicpea (*Cicer*), storage proteins, 21
Chloramphenicol, protein synthesis and, 32
*Chlorella*, nitrate reductase, 82
  maize root proteinase and, 85—86
*p*-Chloromercuribenzoate (pCMB), soybean nodule exopeptidase activity and, 119
Chlorophyll
  in chloroplast senescence, 131—134
  degradation, 139
Chlorophyll-protein complex, 131
  in chloroplast senescence, 132
Chlorophyllase, 139
Chloroplast(s)
  autonomous nature of, 138—142
  polypeptides, 73
    cytoplasmically synthesized, posttranslational processing of, 74—77
  proteins, 70
  proteolysis in, 60
  proteolytic activity, 59
Chloroplast sensescence, 125—126
  characteristics, 126—137
  chlorophyll degradation, 139
  $CO_2$ assimilation during, 134
  compartmentation of proteolysis, 143—145
  composition changes, 131—134
    lipids, 134
    pigment/pigment-protein complexes, 131—134
  degradative events, vacuolar influences, 137—138
  functional changes during, 134—137
  growth regulators and, 131
  induced senescence, 130—131
  initiation of proteolysis in, 145—146
  leaf detachment and, 130—131
  lipids in, 134
    degradation of, 139, 140
  natural, ultrastructural changes, 127—130
  nuclear and chloroplast genomes and, 142—143
  nucleic acid degradation, 139—140
  numerical changes in chloroplast population, 126—127
  photophosphorylation during, 135
  photosynthetic electron transport during, 135
  pigment/pigment-protein complexes in, 131—134
  pinocytosis in, 137—138
  proteases in, 142—146
  protein degradation, 140—142
    light harvesting chlorophyll a/b protein, 140—141
    NADPH-protochlorophyllide oxidoreductase, 141—142
    $Q_\beta$-protein, 141
    RuBPCase, 142
  protein synthesis and, 136—137
  stromal protein degradation, 145
  thylakoid protein degradation, 143—145
  tonoplast transfer properties in, 138
  ultrastructural changes
    induced senescence, 130—131
    natural senescence, 127—130
*Cicer arietinum*, 21
$CO_2$ assimilation, in leaf senescence, 134
Compartmentation of proteases, proteolysis in senescing leaves and, 59—60
α-Conarachin, 23
Concanavalin A, 25, 74
β-Conglycinin, 23, 25
Convicilin, 71
Corn, see Maize
Cotranslational transport, 70

Cotton (*Gossypium*)
   acalin B, 25
   carboxypeptidase, 31
   nitrate reductase protector molecule in, 84
Cowpea (*Vigna*)
   root nodules
      protease activity, 117
      proteolytic activity during senescence, 111, 117
   trypsin inhibitor, 33
Crystalloids, 23
   pumpkin seed, 33
*Curcurbita*, see also Pumpkin; Squash
   *maxima*, 31—32
   *moschata*, 25
Cycloheximide
   chloroplast senescence and, 142
   nitrate reductase synthesis and, 82
Cysteine proteinase, 55
   barley leaves, 86
Cystine, half
   in cereal albumins, 4
   in cereal globulins, 4
   in cereal glutelins, 7
   in cereal prolamins, 6
Cyt f/$b_6$-protein complex, 135

# D

Deamidation, 21
Defoliation, root nodule senescence and, 105
Deuterium oxide, protein degradation and, 59
DFP, see Diisopropylfluorophosphate
Dicots (dicotyledonous plants)
   arylamidases, 31
   carboxypeptidases, 30—31
   endopeptidases, see also Endopeptidases, 28—30
   exopeptidases, 30—32
   proteases, 27—33
   protein bodies, 23
   storage proteins, see also Storage proteins, 22—27
Digalactosyldiacylglycerol
   in chloroplasts, 134
   degradation, 139, 140
Diisopropylfluorophosphate (DFP), 29, 30
   carboxypeptidases and, 52
   soybean nodule exopeptidase activity and, 119
Dipeptidases, 31, 32
Disease (plant), root nodule function and, 105
Dithioerythritol, soybean nodule exopeptidase activity and, 119
DNA
   degradation, in chloroplast senescence, 139—140
   nuclear
      chloroplast senescence and, 142, 143
      transcription of, 145
Doxorubicin, inhibition of IFP-synthase-activating proteinase, 101
Dwarf bean leaves, azocasein degradation, 56

# E

Edestin, 25
Electron transport
   photosynthetic, during senescence, 135
   $Q_\beta$-protein and, 141
Electrophoretic studies
   cereal prolamins, 5
   structural changes in storage proteins during mobilization, 21—22
Endopeptidases (proteinases), 27
   in barley, 11
   barley leaf proteinase, 86
   of dicots, 28—30
   of germinated seeds, 30
   in germinating seeds, 55
   IFP-synthase-activating
      $Ca^{2+}$-dependent generation from *Poterioochromonas* membranes, 97—98, 100
      cell shrinkage and, 98—101
      inhibition of, 100, 101
      purification and properties of, 96—97
   during leaf senescence, 53—57
      annual vs. perennial plants, 53
      changes in activity patterns, 56—57
      changes in total activity, 53—55
   in maize, 12
   maize root proteinase, 85—86
   nitrate reductase inactivation and, 83—84
   nitrate reductase regulation and, 81—87
   protease E, 29
   of quiescent hemp, 29
   in rice, 13
   from senescing tobacco leaves, 54—55
   in sorghum, 13
   in soybean nodules, 115, 116
   in storage protein hydrolysis, 10—13
   of ungerminated (dormant) seeds, 29—30
   in wheat, 11
Endoplasmic reticulum, rough (RER), protein body formation, 2
Endosperm, protein bodies, 2
Enzymes
   proteolytic, see Proteases
   trypsin-like, 29
Ethylene, chloroplast senescence and, 131
Ethylenediaminetetraacetic acid (EDTA), 29, 30
*N*-Ethylmaleimide (NEM), 29
   proteolytic activity and, 54
Exopeptidases (peptidases), 27—28
   alkaline peptidase, 31—32
   arylamidases, 31
   carboxypeptidases, 30—31
   of dicots, 30—32
   root nodules, pH optima, 118
   in senescing leaves, 51—53
   soybean nodule, modifying agents, 119
Euphaseolin, 38

## F

Fatty acids
  free, in chloroplast senescence, 135
  proteolysis and, 60—62
  unsaturated, activation of proteolysis and, 145—146
Favin, 74
*Festuca pratensis* mutant, 60
  chloroplast senescence, 138
    nuclear DNA and, 143
    thylakoid protein degradation, 143
  thylakoid protein degradation during senescence, 133—134
Festucoideae, see Barley; Oats, Rye; Wheat
*Ficus carica*, 55
Flax proteinase, 85

## G

Galactolipase, 139
β-Galactosidase, 139
α-Galactosyl-(1→1)-glycerol, see Isofloridoside
α-Galactosyl-(1→1)-glycerol-3-phosphoric acid synthase, see IFP-synthase
Germination
  dicotyledonous plants, 20
  mobilization of protein reserves during, 1—13
  pumpkin seed, 33—35
  storage protein mobilization (hydrolysis)
    in dicots, 19—40
    in monocots, 6—13
Gibberelic acid, 12
Ginger (*Zingiber*), proteolytic activity, 55
Gliadin, 28
Globoids, 23
Globulins, see also Storage proteins, 23
  of cereals, 3, 4
  of dicots, 23—25
  $E_{\alpha\beta}$, 34
  $F_{\alpha\beta}$, 34
  of legumes, proteolysis of, 35—37
  legumins, 23—25
  pumpkin, 25, 33
  7S, 25
  vetch, proteolysis, 35—37
  vicilins, 23, 25
Glucose-6-phosphate dehydrogenase, inactivation, 61
Glutamate dehydrogenase, during senescence, 51
Glutamate synthase, during leaf senescence, 51
Glutamic acid
  in cereal albumins, 4
  in cereal globulins, 4
  in cereal glutelins, 7
  in cereal prolamins, 6
Glutamine synthetase
  inactivation, 58, 61
  during leaf senescence, 51

Glutelins, see also Storage proteins, 23
  of cereals, 5—7, 72
Glycine
  in cereal albumins, 4
  in cereal globulins, 4
  in cereal glutelins, 7
  in cereal prolamins, 6
*Glycine max*, see also Soybean, 21
  vicilin, 71
Glycinin, 23
  properties, 24
  subunit composition, 24
Glycosylation, legume storage protein, 71
*Gossypium hirsutum*, see also Cotton, 25
Graminae, see also Cereals
  leaves, protein mobilization during senescence, 53—54, 56
Growth regulators, chloroplast senescence and, 131
Guanosine triphosphate (GTP), proteolysis and, 61

## H

Hemoglobinase I, 57
Hemoglobinase II, 57
Hemoglobin substrate, 53
Hemp (*Cannabis*)
  edestrin, 25
  quiescent, endopeptidase, 29
Histidine
  in cereal albumins, 4
  in cereal globulins, 4
  in cereal glutelins, 7
  in cereal prolamins, 6
Hordeae, see Oats
*Hordeum*, see also Barley
  *distichium*, 86
  *vulgare*, 86
Hydrolysis, see also Proteolysis
  by arylamidases, 31
  storage proteins, 6—10
    pathway for, 39—40
    proteolytic enzymes and, 10—13
*p*-Hydroxymercuribenzoate (pHMB), 29
  proteolytic activity and, 54

## I

IFP-synthase, 93
  activation of, 93—96
    activating proteinase, 96—98
    cell shrinkage and, 98—101
  activating proteinase, 96—98
    $Ca^{2+}$-dependent generation from membranes, 97—98, 100
    inhibition of, 100, 101
    purification and properties, 96—97
Immunoaffinity chromatography, storage proteins, 22
Immunoblotting, storage proteins, 22

Inhibitors
    nitrate reductase, 83—84
    proteinase, see Proteinase inhibitors
Iodoacetate, 29
Isofloridoside (IF), 92
    metabolism in *Poterioochromonas*, 93—96
"Isolectin", 26
Isoleucine
    in cereal albumins, 4
    in cereal globulins, 4
    in cereal glutelins, 7
    in cereal prolamins, 6

## J

Jackbean (*Canavalia*)
    endopeptidase, 30
    lectin, 25

## K

Kafirin, hydrolysis, 9
Kidney bean, see also *Phaseolus vulgaris*
    carboxypeptidase inhibition, 32—33
    glycoprotein mobilization, 38
    lectin proteolysis, 38
Kinetin, chloroplast senescence and, 131
KSTI, see Kunitz soybean trypsin inhibitor
Kunitz assay, plant protelytic activity, 28
Kunitz soybean trypsin inhibitor (KSTI), 26—27
    amino acid composition, 24
    properties, 26—27
    proteolysis, 39
Kiwi fruit, 55

## L

*Lactuca sativa*, 29, 32
Latex, endopeptidase in, 55
Leaf senescence, see also Chloroplast senescence
    aminopeptidases in, 51—52
    artificial, 51
    carboxypeptidases in, 52—53
    chlorophyll degradation, 139
    $CO_2$ assimilation during, 134
    endopeptidases in, 53—57
        changes in activity patterns, 56—57
        changes in total activity, 53—55
    exopeptidase activity in, 51—53
    metabolic changes during, 50
    natural, 50—51
    nitrogen metabolism during, 50—51
    protein degradation during, 50—63
    proteolysis regulation in, 57—62
        compartmentation and, 59—60
        low molecular weight compounds and, 60—62
        pH and, 60
        protease levels and, 57—58

        susceptibility of substrate proteins and, 58—59
    proteolytic enzymes in, 49—63
Lectins, 25—26
    amino acid composition, 24
    processing of, 73—74
        properties, 26
        proteolysis, 38
        subunit composition, 26
β-Lectins, 26
Leghemoglobin
    alfalfa nodule, 121
    pea nodule, 122
    reduction in, 105
Legumes
    leaf senescence, endopeptidase activity in, 54, 56—57
    lectin and trypsin inhibitor proteolysis, 38—39
    root nodules, 103—122
        aging, 105
        ontogeny, 104
        elongate cylindrical, 104, 106—109, 118, 120—122
        senescence, 104—105
        spherical, 104, 105, 110—118
    storage globulins, 23—25
        proteolysis, 35—37
    storage proteins, 71
        mobilization of, 35—39
Legumins, 23—25, 71
    peanut, 23
    properties, 23—24
    proteolysis, 35—37
    soybeans, 23
*Lemna minor*, 59
    change in tonoplast transfer properties, 138
Lettuce (*Lactuca*)
    proteolytic activity, 29
    trypsin-like enzyme in, 32
Leucine
    in cereal albumins, 4
    in cereal globulins, 4
    in cereal glutelins, 7
    in cereal prolamins, 6
Leucine aminopeptidase, 31
Leupeptin, 58
    nitrate reductase degradation and, 86
Light harvesting chlorophyll a/b protein, degradation, 140—141
Lipids, 20
    chloroplast, 134
        degradation, 139, 140
Low molecular weight compounds, proteolysis in senescing leaves and, 60—62
Lupin (*Lupinus*), endopeptidase, 28
*Lupinus angustifolius*, 28
Lysine
    in cereal albumins, 4
    in cereal globulins, 4
    in cereal glutelins, 7
    in cereal prolamins, 6

## M

Maize
  albumins, 3, 4
  globulins, 3, 4
  glutelins, 6, 7
  leaf
    nitrate reductase, 82
    RuBPCase, 50
  prolamins, 3, 5, 6
  protein body formation in, 2
  proteolytic enzymes in, 12
  root, proteinase, 83—87
  scutellum, nitrate reductase activator, 84
  storage protein hydrolysis in, 9—10
    proteases and, 12—13
Membrane proteins, 60, 62, 70
Metabolism
  during leaf senescence, 50
    nitrogen, 50—51
  isofloridoside, 93—96
Methionine
  in cereal albumins, 4
  in cereal globulins, 4
  in cereal glutelins, 7
  in cereal prolamins, 6
Mitochondria, proteins, 70
Molecular weight
  arylamidases, 31
  Bowman-Birk soybean trypsin inhibitor, 26
  convicilin, 71
  glycinin, 24
  Kunitz soybean trypsin inhibitor, 26
  lectins, 26
  legumins, 23
  vicilins, 25
Monocots, see Cereals
Monogalactosyldiacylglycerol
  in chloroplasts, 134
  degradation, 139, 140
Mung bean
  arylamidase, 31
  trypsin inhibitor, degradation of, 38—39
  vicilin, 25, 37
  vicilin peptidohydrolase, 32
Mustard seedlings, leaf senescence, 131

## N

NADH, nitrate reductase inactivation and, 84
NADPH-protochlorophyllide oxidoreductase, 141—142
*Neurospora*, nitrate reductase, 82
Nitrate, root nodule funciton and, 105
Nitrate reductase
  activity, 82
  degradation by plant proteinases, 85—87
    by barley leaf proteinase, 86
    by maize root proteinase, 85—86
  half-life, 82
  inactivation, 58
    pyridine nucleotides and, 61
    reversible, 84
  inactivators, 83—84
  inhibitors, 83—84
  in leaf senescence, 51
  protector molecules, 84
  proteinase inactivators, 83, 84
  regulation of, endopeptidases in, 81—87
  reversible inactivation by NADH, 84
  structure, 82
  turnover rate, 82—83
Nitrogen metabolism, during leaf senescence, 50—51
Nodules, see Legumes, root nodules
Nutrient reserves, dicotyledonous plants, 10

## O

Oats
  albumins, 3, 4
  globulins, 3, 4
  glutelins, 6, 7
  prolamins, 5, 6
  storage protein hydrolysis in, 8—9
    proteases and, 12
*Ochromonas,* see *Poterioochromonas*
Oligosaccharides, 20
*Oryza sativa,* see also Rice, 74
Oryzeae, see Rice
Orzoideae, see Rice
Osmotic solutes, 92
Osmotic systems, 92

## P

P20 (small subunit precursor), in study of chloroplast processing activity, 75—77
Palmitoyl-DL-carnitine, inhibition of IFP-synthase-activating proteinase, 101
Panecoideae, see Maize; Sorghum
Papaya (*Carica*), 55
Pea (*Pisum*)
  chloroplast $Q_\beta$-protein, 141
  leaf senescence, 54
  lectin, 74
  root nodule, leghemoglobin and bacteroid number during dark-induced senescence, 122
  storage proteins, 21
  vicilin, 25, 72
Peanut (*Arachis*)
  carboxypeptidase, 30
  α-conarachin (arachin), 23
  protein bodies, 23
  storage proteins, 21
Peptidases, see Exopeptidases
Peptide hydrolase, see Proteases
Peroxidase, 130

pH
    endopeptidase in senescing leaves and, 53—56
    of nodule exopeptidases, 118
    proteolysis in senescing leaves and, 60
Phaseolin, processing, 72
*Phaseolus vulgaris,* see also Kidney bean, 21, 29, 56
    leaf senescence, 133
        electron transport during, 135
    legumin/vicilin ratio, 71
    vicilin, 71
1,10-Phenanthroline, 29, 30
    soybean nodule exopeptidase activity and, 119
Phenylalanine
    in cereal albumins, 4
    in cereal globulins, 4
    in cereal glutelins, 7
    in cereal prolamins, 6
Phenylmethylsulfonylfluoride (PMSF), 29, 30
    carboxypeptidases and, 52, 54
Photogene 32, 141
Photophosphorylation, during senescence, 135
Photosynthesis, 126
Phytins, 20
Phytohemagglutinins, see Lectins
Pigments, in chloroplast senescence, 131—134
Pineapple (*Ananas*), endopeptidase activity, 55
Pinocytosis, in chloroplast senescence, 137—138
*Pisum sativum,* see also Pea, 21, 71
Plastidquinones, lipophilic, in plastoglobuli, 127
Plastoglobuli, in chloroplast senescence, 127, 130
PMSF, see Phenylmethylsulfonylfluoride
Polyacrylamide gel electrophoresis (PAGE), storage proteins, 21—22
Polyamines, proteolysis in senescing leaves and, 60
Polymyxin B, inhibition of IFP-synthase-activating proteinase, 101
Posttranslational modification
    cytoplasmically synthesized chloroplast polypeptides, 74—77
    lectins, 73—74
    storage proteins, 70—73
*Poterioochromonas malhamensis,* 92
    cell volume and turgor pressure regulation in, 92—93
    IFP-synthase activation in, 93—96
        $Ca^{2+}$-dependent generation of proteinase from membranes, 97—98
        cell shrinkage and, 98—101
        purification and properties of activating proteinase, 96—97
    isofloridoside metabolism in, 93—96
Precursor proteins, 72
Preproproteins, 72
Prolamins, see also Storage proteins, 23
    of cereals, 3—6, 72
Proline
    in cereal albumins, 4
    in cereal globulins, 4
    in cereal glutelins, 7
    in cereal prolamins, 6

    formation of, 12
Protease I, of pumpkin, 29, 34, 35
Protease II, of pumpkin, 34, 35
Proteases
    A, of vetch, 36
    activity levels
        proteolysis in senescing leaves, 57—58
        in senescing leaves, 51—57
    B, of vetch, 36
    in barley, 11—12
    C, of vetch, 36
    in chloroplast senescence, 142—146
    classification, 27—28
    compartmentation, proteolysis in senescing leaves and, 59—60
    D, of vetch, 36
    of dicots, 27—33
    E, 29
    of vetch, 36
    elongate nodule senescence and, 118, 120—122
    in leaf senescence, see also Leaf senescence, 49—63
    legume root nodules
        elongate nodules, 118, 120—122
        spherical nodules, 105, 111—118
    in posttranslational modification of proteins, 69—77
    regulation by endogenous proteinase inhibitors, 32—33
    in rice, 13
    of seeds, time course studies, 20—21
    in sorghum, 13
    spherical nodule senescence and, 111—118
    in storage protein mobilization (hydrolysis)
        in dicots, 19—40
        in monocots, 10—13
        in wheat, 10—11
Protein(s)
    degradation, see also Proteolysis
        in chloroplast senescence, 140—142
        during leaf senescence, 50—63
    half-lives, in cereal leaves, 50
    leaf, localization, 59
    light harvesting chlorophyll a/b protein, 140—141
    membrane, 60, 62
    mobilization, in senescing Gramineae leaves, 53—54
    posttranslational modification, 69—77
    $Q_\beta$, 141
    reserves, see Storage proteins
    transport, 70
    11S, 24
Proteinase inhibitors, 26—27
    interaction with trypsin inhibitors, 32, 33
    regulation of proteolytic enzymes and, 32—33
    physiologic function, 27
    pumpkin arylamidase, 35
Protein bodies
    dicots, 23
    endoprotease, 74
    formation during maturation, 2—3

pumpkin seed, 33
Proteolysis
  activation by unsaturated fatty acids, 145—146
  ATP-dependent, 62
  in chloroplast senescence
    compartmentation of, 143—145
    initiation of, 145—146
  compartmentation of, 143—145
  electrophoretic mobility of proteins and, 21
  legume lectins and trypsin inhibitors, 38—39
  legume storage globulins, 35—37
  posttranslational, 70
  pumpkin storage proteins, 33—35
  in senescing leaves
    compartmentation and, 59—60
    low molecular weight compounds and, 60—62
    pH and, 60
    protease levels and, 57—58
    regulation of, 57—62
    substrate susceptibility and, 58—59
  in senescing soybean nodules, 111
Protochlorophyllide, 141
Pumpkin (*Curcurbita*), 25, 31—32
  arylamidase, 35
  endopeptidase, 28, 29
  germination, 33—35
  globulin, 25, 33
  protease I, 29, 34, 35
  protease II, 34, 35
  storage protein mobilization, 33—35
Pyridine nucleotides, enzyme inactivation and, 61

## Q

$Q_\beta$-protein, degradation, 141

## R

Raffinose, 20
Ribulose biphosphate carboxylase (RuBPCase)
  in cereal leaves, 50
  synthesis, in chloroplast, 136—137
  turnover, 142
Rice (*Oryza*)
  albumins, 3, 4
  globulins, 4
  glutelins, 6, 7
  lectin, 74
  prolamins, 5, 6
  proteases, 13
  storage protein hydrolysis in, 9
    proteases and, 13
Rice cell nitrate reductase inhibitor, 83—84
Ricin, 74
*Ricinus communis*, 27, 30
  agglutinin, 74
RNA
  chloroplast, 137
  degradation, in chloroplast senescence, 139—140
RuBPCase, see Ribulose biphosphate carboxylase
Rye
  albumins, 3, 4
  globulins, 3, 4
  glutelins, 6, 7
  prolamins, 5, 6
  storage protein hydrolysis in, 7—8
    proteases and, 12

## S

Scutellum, 7
SDS-PAGE
  pumpkin globulin, 34
  storage proteins, 22
Sedimentation coefficient
  legumins, 23
  vicilins, 25
Seeds, germinating, endopeptidases in, 55
Senescence
  chloroplasts, see also Chloroplast senescence, 125—146
  dark-induced, root nodules and, 122
  leaves, see Leaf senescence
  legume root nodules, 104—105
    elongate nodules, 118, 120—122
    proteases and, 105, 111—122
    spherical nodules, 105, 111—118
Serine
  in cereal albumins, 4
  in cereal globulins, 4
  in cereal glutelins, 7
  in cereal prolamins, 6
Serine proteinase
  maize root, 87
  wheat leaf, 85
Shading
  leghemoglobin content and, 105
  senescence and, 105
Solanaceae, 54—55
Sorghum
  albumins, 3, 4
  globulins, 4
  glutelins, 6, 7
  prolamins, 3, 5, 6
  storage protein hydrolysis in, 9—10
    proteases and, 13
Soybean, see also *Glycine max*
  aspartic proteinase, 29
  carboxypeptidase, 30
  β-conglycinin (glycinin), 23, 25
  leaves
    azocollases, 54, 56
    hemoglobinases, 57
    nitrate reductase inhibitor, 83, 84
  root nodules
    endopeptidase activity, 114—116
    exopeptidase activity, 115, 119
    proteolytic activity during senescence, 111
    shading/darkness and, 105

storage proteins, 21
    trypsin inhibitors, 26—27
Spermatophytes, 20
Spermidine, proteolysis and, 60
Spermine, proteolysis and, 60
Spinach leaf, nitrate reductase, 82
Squash (*Curcurbita*)
    dipeptidase, 32
    leucine aminopeptidase, 31—32
Stachyose, 20
Starch, 20
Storage proteins, see also specific protein group
    albumins, 3, 22, 27
    classification, 3
    of dicots, 22—27
    globulins, 3, 23—25
    glutelins, 3, 5—7, 23
    mobilization (hydrolysis)
        in germinating dicots, 19—40
        in germinating monocots, 6—13
        in legumes, 35—39
        pathway for, 39—40
        proteases in, 10—13
        in pumpkin, 33—35
        structural changes during, 21—22
        study approaches, 20—22
        time course studies, 20—21
    nature of, 2—7
    processing of, 70—73
    prolamins, 3—5, 23
    properties of, 3—6
    protein body formation during maturation, 2—3
Stromal protein, degradation, in chloroplast senescence, 145
Substrate
    for endopeptidase assay, 28—29
    protein, targeting of, 145
    susceptibility, and proteolysis in senescing leaves, 58—59
    in time course studies of seed proteases, 21

## T

Temperature stress, root nodule, 105
Threonine
    in cereal albumins, 4
    in cereal globulins, 4
    in cereal glutelins, 7
    in cereal prolamins, 6
Thylakoids
    in chloroplast senescence, 127
    lipid degradation, 144
    protein degradation, 132—134, 143—145
Time course studies, proteases in protein mobilization, 20—21
Tobacco (Solanaceae) leaf, proteolytic activity, 54—55
Tomato leaves, carboxypeptidase activity, 52
Tonoplast, transfer properties, senescence and, 138

Tree leaves, proteolytic activity during senescence, 55
Triazene herbicides, $Q_\beta$-protein and, 141
Trichloroacetic acid, in endopeptidase activity, 28
Triglycerides, 20
*Triticum*, germination, storage protein hydrolysis during, see also Wheat, 8
Trypsin inhibitors
    Bowman-Birk, 26
    interaction with proteinase inhibitors, 32, 33
    Kunitz, 26
    proteolysis, 38—39
Tubulin, hydrolysis, 61
Tungstate, nitrate reductase synthesis and, 82
Turgor pressure, regulation of, in *Poterioochromonas*, 92
Tyrosine
    in cereal albumins, 4
    in cereal globulins, 4
    in cereal glutelins, 7
    in cereal prolamins, 6

## U

Ubiquitin, ATP-dependent proteolysis and, 62

## V

Vacuole
    in chloroplast senescence, 137—138
    endopeptidase in, 55
Valine
    in cereal albumins, 4
    in cereal globulins, 4
    in cereal glutelins, 7
    in cereal prolamins, 6
*Valonia*, 92
Vetch (*Vicia*)
    endopeptidase, 28
    storage globulin proteolysis, 35—37
    storage proteins, 21
    ungerminated, endopeptidase, 29
*Vicia*
    *faba*, 23, 71
    *sativa*, see Vetch
Vicilin(s), 23, 25, 71
    amino acid composition, 24
    euphaseolin, 38
    inhibitor, 32
    processing, 72—73
    properties, 25
    proteolysis, 37
    subunit composition, 25
Vicilin peptidohydrolase, 32, 37
*Vignas unguiculata*, see also Cowpea, 33

## W

Water stress, root nodule, 105

Wheat
- albumins, 3, 4
- globulins, 4
- glutelins, 6, 7
- leaves
  - aminopeptidase activity, 51—52
  - carboxypeptidase activity, 52—53
  - caseolytic activity, 58
  - endopeptidase activity, 53, 54
  - glutamine synthetase activity, 58, 61
  - nitrate reductase protector molecules, 84
  - nitrate reductase inactivator, 83, 84
  - nitrogen mobilization, 52
- prolamins, 3, 5, 6
- proteases in, 10—11
- protein body formation in, 2
- storage protein hydrolysis in, 7—8
  - proteases and, 10—11
- thylakoid protein degradation, 144

## Z

*Zea mays*, RuBPCase turnover, see also Maize, 142
Zein, 72
- hydrolysis, 13

*Zingiber*, 55